Florida Hurricanes and Tropical Storms

1871-1993: An Historical Survey

Fred Doehring, Iver W. Duedall, and John M. Williams

Division of Marine and Environmental Systems,
Florida Institute of Technology
Melbourne, FL 32´

Technical Paper

June 1994

$5.00

Copies may be obtained from:

Florida Sea Grant College Program
University of Florida
Building 803
P.O. Box 110409
Gainesville, FL 32611-0409
904-392-2801

Our friend and colleague, Fred Doehring pictured below, died on January 5, 1993, before this manuscript was completed.

Until his death, Fred had spent the last 18 months painstakingly researching data for this book. Fred had a genuine interest in Florida Tech, and in helping students find information on weather.

We thoroughly enjoyed working with Fred and we are hopeful that this book will enhance hurricane awareness for all Florida residents.

We dedicate this book to his family.

Iver W. Duedall

John M.Williams

The hurricane as a heat engine, is inefficient, hard to start, and hard to sustain; but once set in motion, once mature, is an awesome natural event!

From birth, the hurricane lives in an environment that constantly tries to kill it...... and ultimately succeeds

Dr. Robert C. Sheets
Director
National Hurricane Center
1990

Table of Contents

List of Tables, Figures, and Plates

Hurricane and Tropical Storm Tracks by 10-year periods

Preface

This book presents, by historical periods, a summary of the hurricanes and tropical storms that struck Florida's more than 1200-mile long coastline during the 122 years from 1871 through 1993.

Prior to the publication of this book, the only books or reports exclusively with Florida hurricanes were R.W. Gray's report, revised by Grady Norton in 1949, titled Florida Hurricanes, and a survey by Gordon E. Dunn and staff of the National Hurricane Center (NHC) also titled Florida Hurricanes.

Grady Norton and Gordon Dunn were important figures during the early creation of what is now known as the National Hurricane Center. Grady Norton was considered by many as one of the best hurricane forecasters. After Grady Norton's death in 1954, Gordon Dunn was named director of the NHC. After Gordon Dunn retired, Dr. Robert Simpson became NHC Director. He was followed by Dr. Neil Frank. The current NHC Director is Dr. Robert C. Sheets.

In addition to the report Florida Hurricanes, the very recent 1992 NHC report titled The Deadliest, Costliest, and Most Intense United States Hurricanes of this Century provides invaluable information on both historical and recent hurricanes affecting Florida and the United States.

Our primary goal in preparing this book was to update the historical work as it pertains to Florida, to consolidate and standardize technical terms published at the beginning of each hurricane season on hurricane tracking maps, and to introduce the following new material pertaining to Florida: (1) a detailed historical discussion, (2) a chronological listing of all Florida hurricanes, (3) 13 plates of hurricane and tropical storm tracks grouped into 10-year increments, and (4) a table showing the number of tropical storms and hurricanes by 10-year increments. The book is written on a non-technical level for the general reader who is interested in knowing when and where hurricanes affected Florida and the magnitude of damage inflicted by the storms. Those wishing more technical information on hurricanes can consult the references or contact the NHC directly.

Principal reference documents used in preparing this book, in addition to the ones mentioned above, came from the National Oceanic and Atmospheric Administration (NOAA) publication,

Historical Climatology Series 6-2, Tropical Cyclones of the North Atlantic Ocean, 1871-1986, and U.S. Weather Bureau publications, Climatological Data for Florida 1897-1965.

It should be noted that this book deals primarily with the characteristics of Florida Hurricanes and some eyewitness accounts. Amounts of precipitation associated with Florida hurricanes were not discussed in this book because of their frequent occurrence resulting from other systems such as tropical depressions or non-tropical systems. However, we should point out that precipitation from a hurricane can be very great and can cause major damage and flooding.

While tropical depressions can cause heavy flooding problems and damage, rainfall from tropical depressions is not discussed. Also, tropical waves and depressions are not named either.

We should mention that some of the direct quotes we use make reference to the earlier used term "Great Hurricane" and "Major Hurricane". The reader is referred to the glossary and tables for a detailed explanation of all terms, including the Saffir-Simpson Scale now in use to categorize hurricanes. In some instances, we have made inserts, indicated by [...], into quotations to provide clarification. The [...] notation was also used to provide an estimate of the damage in dollars, adjusted to 1990.

We are especially thankful to the reviewers of the original manuscript who pointed out several deficiencies and errors in the first writing and who provided important suggestions leading to an improved and accurate final manuscript and to friends, colleagues, assistants, and organizations whose help we could not have done without. Specifically, we thank Bill Mahan who encouraged us to prepare this book, Annette Bernard, Ann Bergonzoni, Derrick Doehring, Rosary Pedreira, Arnold Samreth, and Huan Feng for manuscript preparation, to Henry (Hank) Brandli for graciously providing us with his satellite images of Florida Hurricanes, to Rob Downey for the color photograph of Hurricane Andrew, to Anita Bromberg and John Reposa who assisted in the preparation of the plates showing the hurricane tracks, Victoria (Tori) Smith and Jeanette C. Sparks of the Florida Tech Evans Library for searching historical documents, to the Melbourne Office of the National Weather Service for providing the Doppler radar image of Hurricane Andrew, to Florida Sea Grant College who provided financial assistance to complete this work and to Jay Humphreys who read the manuscript and provided suggestions for improvement and to Susan Grantham for the layout and design, to the News/Sun Sentinel, Ft. Lauderdale, and the Miami Herald, Miami, for the use of their photographs, to Ms. Ruth Warner for kindly providing us with her grandmother's account of the 1926 Miami hurricane, to Lois Stephens for allowing us to use her personal account of Hurricane Andrew entitled "On Sabbatical with Hurricane Andrew", and to the National Hurricane Center for providing photographs.

Chapter 1

Introduction

A hurricane is an extremely violent whirling and spiraling tropical cyclone, shaped somewhat like a funnel, that frequently originates in tropical regions of the North Atlantic Ocean, Caribbean Sea, Gulf of Mexico and eastern North Pacific Ocean. The term cyclone, used by weathermen and meteorologists, refers to an area of low pressure in which winds move around the pressure center and are usually attended by foul weather and strong wind speeds. A tropical cyclone is a nonfrontal, warm-core, low pressure synoptic scale storm that develops over tropical or subtropical waters and has a definite organized circulation.

Tropical Cyclones are called typhoons in the western North Pacific Ocean, hurricanes in the eastern north Pacific, baguios in the South China Sea, cyclones in the Indian Ocean, and willy-willies in Australia.

The size of a typical hurricane can vary considerably depending on the extent of the tropical storm's wind fields and rain fields. In a relatively large hurricane, such as the Florida hurricane of September 1947, hurricane force winds can extend 100 miles from the center (i.e. a distance from Palm Beach to Melbourne). However, in August 1992 Hurricane Andrew, the most destructive hurricane ever to strike Florida, or the U.S. mainland for that matter, had maximum winds with a radius of only about 12.5 miles. Thus hurricanes vary considerably in their size.

To be classified officially as a hurricane, wind speed in a tropical cyclone must be 74 miles per hour or greater. The direction of rotation of wind in a hurricane is counterclockwise in the northern hemisphere, and clockwise in the southern hemisphere. The average hurricane's center, referred to as the eye, is about 14 miles in diameter. The eye is surrounded by hurricane force winds, and is known as the wall cloud, or eye wall. Outside the wall cloud, or area of maximum winds, winds decrease fairly rapidly to tropical storm or gale force.

Within the hurricane, barometric pressure is 1-3 inches of mercury below the standard atmospheric pressure at sea level which is 29.92 inches of mercury.

The North Atlantic hurricane season occurs during the months of

June through November, with September generally having the largest number. The total number of hurricanes or tropical storms show great variation from year to year. In fact, certain past multi-decade periods had significantly greater numbers of hurricanes than others. This is supported by the recent study of William M. Gray in 1990 who reported that the period from the late 1940s through the late 1960s had a much larger number of hurricanes (i.e. a strong cycle) than during the 1970s and 1980s, except for 1988 and 1989 (i.e. a weak cycle).

Quoting a passage from Gray's 1990 article entitled *Strong Association Between West African Rainfall and U.S. Landfall of Intense Hurricanes:*

> Seasonal and multi-decadal variations of intense hurricane activity are closely linked to seasonal and multi-decadal variations of summer rainfall amounts in the Western Sahel region of West Africa.
>
> In general, the annual frequency of intense Atlantic hurricanes was appreciably greater from 1947 to 1969, when plentiful amounts of rainfall occurred in West Africa, than during the years between 1970 to 1987, when drought conditions prevailed.

The average forward movement of a hurricane approaching the Florida coastline is about 6-15 miles per hour. The direction of hurricane movement relative to the coastline has a large bearing on added destructive forces, with the perpendicular landfall of a hurricane being the most dangerous situation. This is because the wind field in a hurricane is typically asymmetric with the strongest wind generally within the right-front quadrant of the storm as viewed from the direction of movement and with the forward speed added to the wind speed. The right-front quadrant is the side of the wind field which produces the strongest storm surge, which, in most cases, is the most destructive part of the hurricane.

A storm surge, also called a hurricane surge, is the abnormal rise in sea level accompanying a hurricane or any other intense storm. The height of the storm surge is the difference between the observed level of the sea surface and the level in the absence of the storm. The storm surge is estimated by subtracting the normal or astronomical tide from the observed or estimated storm tide. Surge heights vary considerably and result from a combination of direct winds and atmospheric pressure. Water transport by waves and swells, rainfall, and shoreline configuration, bottom topography, and tide heights at the time the storm or hurricane hits the coast are also factors. As an example of an extreme storm surge, Hurricane Donna which struck the Florida Keys in 1960, caused a surge of an estimated 12 to 14 feet, which is very significant considering the fact that there are few structures and little terrain that high in the Keys. A more catastrophic surge was the 24.4 foot surge which resulted from Hurricane Camille which struck the Mississippi coastline in 1969. The potentially devastating effects of the storm surge are further illustrated if one considers that a cubic yard of seawater weighs nearly three-fourths of a ton which pretty well guarantees destruction of anything in its path.

The storm surge has a tendency to dissipate the farther inland it goes, particularly if the land rises in elevation. However, winds and some degree of flooding are still remaining problems. High winds, the storm surge, battering waves, and high tide make a hurricane a potentially deadly killer with accompanying devastation and huge losses to property. In addition, tornadoes can be spawned by hurricanes, adding to the overall threat.

Torrential rainfall, which can also occur in a hurricane, adds to life-threatening and major damaging effects of a hurricane by causing floods and flash floods. For example, the aftermath of Hurricane Agnes, which was a relatively weak Florida hurricane, but well known as one of the costliest hurricanes in the mid-Atlantic states, resulted in severe inland flooding from torrential rainfall from its merging with another weather system in mountainous areas. In this case the hurricane surge had little part in the destruction that resulted.

High winds alone can lead to a barrage of flying debris, including tree limbs and branches, signs and sign posts, roofing (including entire roofs in major storms), and metal siding, all of which can move through the air like missiles.

Except when crossing completely flat, wet areas, such as extreme south Florida, hurricanes usually weaken rapidly as they move inland. However, the remnants of a hurricane can bring 6 to 12 inches of rain or more to an area as the storm passes. Should a weakened hurricane on land return to the sea, it can regain strength.

It is clear then that entire communities, including residential and business buildings, can be wiped out by a hurricane.

Because of the difficulty in relating the different and varying factors or characteristics of a hurricane to the destruction, the Saffir-Simpson Scale was conceived in 1972 and introduced to the public in 1975 (Simpson and Riehl, 1981). This scale, named in behalf of Herbert Saffir and Robert Simpson, has been used for 20 years to estimate the relative damage potential of a hurricane due to wind and storm surge. The Saffir-Simpson Scale categorizes a hurricane as being either a one, two, three, four or five, depending upon the barometric pressure, the wind speed, and the storm surge (Table 1). A Category 1 hurricane would inflict minimal damage, for example, primarily to shrubbery, trees, foliage, unanchored structures, small craft, and low lying areas which could become flooded. A Category 5 hurricane would cause catastrophic damage such as blown down trees and power lines and poles, overturned vehicles, torn down or blown away buildings, complete destruction of mobile or manufactured homes and in certain instances entire mobile home parks, and massive flooding. For the first time after Hurricane Andrew, the Fujita Tornado Scale was used to assess damage. Dr. Theodore Fujita is an expert on tornadoes and severe weather. F1 to F5 indicates winds from 73 MPH to over 261 MPH.

The practical usefulness of the Saffir/Simpson Scale is that it relates properties of the hurricane to previously observed damage. Historically and before the Saffir/Simpson Scale was developed, hurricanes were referred to as either Great Hurricanes, or Minimal, Major, or Extreme Hurricanes; because these terms are no longer used, the reader is referred to the glossary for an explanation of the these historical terms and to Tables 1 and 3. Tropical storms are named but are not assigned a Saffir/Simpson category number.

Chapter 2

Historical Discussion of Florida Hurricanes

While Florida is often considered synonymous with sunshine and is frequently called the Sunshine State, mention of the state also brings to mind summer or fall tropical storms and hurricanes. These hurricanes move in a west to northwest direction through the Caribbean and Atlantic toward Florida's coast. From the year 1493 to 1870, the Caribbean area and Florida experienced nearly 400 hurricanes as reported by Professor E.B. Garriot in 1900 in his classic study *West Indian Hurricanes*. Many Spanish galleons loaded with gold, silver and other treasure must have met a swift and untimely demise at the hand of a hurricane or tropical storm. As a result, today treasure hunting is an active and frequently profitable business in Florida.

In recent times, from 1871-1993, nearly 1000 tropical cyclones of tropical storm or hurricane intensity have occurred in the North Atlantic, Caribbean Sea, and Gulf of Mexico. Of this total, about 180 have reached Florida, with 75 of these known to have hurricane force winds (wind speed $\geq$ 74 mph) and 105 with tropical storm force winds (39 mph - 73 mph).

During the early 15-year period from 1871 to 1885, there were 30 tropical cyclones of unknown intensity (shown by the solid line on Plates 1 and 2). Historical data indicate that some of these were hurricanes. Because these hurricanes have not been officially documented, they are listed as total combined storms for the purposes of overall count of Florida hurricanes (Table 2).

In the last 122-year period, there were as many as 21 (in 1933) hurricanes and tropical storms during an individual year, and there were 28 years during which no tropical cyclones made landfall or their center passed immediately offshore of the Florida coastline (Fernandina Beach to Key West to Pensacola).

While early records are fragmentary and incomplete, the following is a discussion of the more formidable Florida hurricanes. For convenience and to provide readable hurricane tracks, the discussion examines hurricanes occurring within 30-year periods, divided into 10-year sections. When possible the Saffir/Simpson Scale (Table 1) describes the hurricane category for both past hurricanes (before the scale was developed), and recent hurricanes.

The Early Years, 1871-1900 [1]

Starting in 1871, only a few years after the Civil War, tropical cyclone data became part of the historical inventory of the U.S. Signal Service and later the U.S. Department of Agriculture Weather Bureau which collected, archived and published these data. Relying on early works of authors, such as *West Indian Hurricanes* (Garriott, 1900), annual tropical cyclone tracks for the years 1871-1990 were later published in the NOAA Historical Climatology Series 6-2, *Tropical Cyclones of the North Atlantic, 1871-1986.* The yearly tracks were extracted from that NOAA publication and are presented here by 10-year periods.

Looking at the first 10 years of tropical cyclone tracks (Plate 1), the most striking feature is that only four tropical cyclones entered Florida's coast from the east, southeast, Atlantic, or Caribbean. In contrast, 17 tropical cyclones entered the west coast and panhandle region from the southwest, the northwestern Caribbean and Gulf of Mexico.

The periods from 1881-1890 (Plate 2) and 1891-1900 (Plate 3) show essentially the same pattern except that the concentration of northeasterly tracks shifts further to the south.

We shall see from an examination of the other plates that this pattern changed after the turn of the century. Principally all of the storms which entered the West Coast of Florida came from the northwestern Caribbean or the southern portion of the Gulf of Mexico. No real explanation can be found for this high frequency.

There are some contradictory events reported during these early years which deserve discussing. They pertain to the hurricanes of 1876, 1880, and 1885. In an interview published June 4, 1978, in the *Florida Today* newspaper, the then National Hurricane Center Director Dr. Neil Frank said:

> In [August] 1871 the center of a hurricane slammed into Central

---

[1] Plates 1-3.

> Florida near Cocoa Beach.... In [September/October] 1873 a major Hurricane exited Florida near Melbourne. ....In [August/September] 1880 another major hurricane battered the coast south of Cocoa Beach.

In reference to the 1880 hurricane that "battered the coast south of Cocoa Beach", this hurricane was classified in Richard W. Gray's *Florida Hurricanes* (Revised Edition) as a Great Hurricane. According to his notes, it affected the Palm Beach—Lake Okeechobee section of Florida; nothing is said about Cocoa Beach, but Dunn and Miller in their book *Atlantic Hurricanes* published in 1964 said that the hurricane affected Vero Beach. However, the 1880 track as extracted from NOAA's (1987) Historical Climatology Series 6-2, shows a hurricane entering the East Coast near Cocoa Beach. The area affected by this hurricane could not have been the Palm Beach-Lake Okeechobee section if the hurricane entered the Florida east coast from the east-south-east near Cocoa Beach. If, on the other hand, R. Gray is correct in his finding, then the 1880 hurricane track reported by NOAA has to be a considerable distance south of Cocoa Beach; this contention is amply supported by the hurricane track of August 26-31, 1880, reported by Garriott in 1900 in his book *West Indian Hurricanes* and by the August 1880 track published by Tannehill in 1938 in his book.

In reference to the August 1885 hurricane, the track published by NOAA (1987) along the east coast may also be in error in that the published track is at least 20 miles offshore. According to B. Rabac's (1986) book *The City of Cocoa Beach*:

> The hurricane that hit in 1885 discouraged further settlement. The storm pushed the ocean waves over the barrier island (elevation 10 feet), flooding out the homesteaders. The beach near the Canaveral Light House was severely eroded, prompting President Cleveland and the Congress to allot money for an effort to move the tower one mile west.

The fact that President Cleveland was in office from 1885-1888 provides further support that this was the year of occurrence. It is certainly possible that the 1885 northerly tropical cyclone track shown over the ocean along the Florida East Coast on the NOAA (1987) track chart was slightly displaced (from the correct position), and that the eye of the hurricane actually passed Cocoa Beach. In fact the report by Sugg, Pardue and Carrodus in 1971 shows the 1885 track passed the central East Coast.

The final controversy concerns the hurricane of 1876. Historical information from G.W. Holmes in a letter to a friend in 1876 indicates that the eye of a terrible hurricane passed over Eau Gallie (now part of Melbourne) on the Indian River on a northerly course during the early morning (no date or month was given) of 1876. Dr. Holmes is quoted as follows:

> The wind came from the east at over a hundred miles an hour until about 3:30 AM. The vortex [the eye] came on us for about four hours, during which not a leaf stirred. We began to look for our boats when all at once with a tremendous roar the wind came from the west, with equal violence in the early part of the night.

The quotation implies that the hurricane traveled north along the Indian River or beaches. NOAA (1987) shows a northerly hurricane track for the year 1876, about 30-40 miles east of the coast passing Cape Canaveral during September 12-19, 1876. The 1876 hurricane could easily have been off by 30 miles which brings the eye over Melbourne, and makes the effect which Mr. Holmes quotes very valid. In October 12-22, another hurricane exited near West Palm Beach from the west. However, until hurricane tracks for 1876, 1880 and 1885 are officially modified by NOAA, they have to be accepted as given from NOAA's track book and shown in Plate 1.

Beginning with the year 1886, tropical storm and hurricane tracks were published separately. In this report, they are presented by dashed and solid lines with the year circled at the beginning of each track (Plates 2 and 3). A solid line prior to 1886 indicates either a tropical storm or hurricane. From 1886 on, a solid line crossing the coast indicates a hurricane, and a dashed line indicates a tropical storm. Beginning with the year 1899, tracks became more detailed and categories were used to describe the relative magnitude of hurricanes.

We conclude this section with quotations about two hurricanes which entered Florida in 1898 and 1899.

Hurricane of October 2-3, 1898, Fernandina Beach

The damage to Fernandina and vicinity was very great. It is conservatively estimated at $500,000. Nothing escaped damage and a great deal was absolutely destroyed. Giant oaks were snapped off at the base, houses blown down, and vessels swept inland by an irresistible in-rush of water. The wind signal display man Major W.B.C. Duryee, who has resided in Fernandina more than thirty years, states that no previous storm was so severe (U.S. Weather Bureau, October 1898).

In 1898, Professor F.H. Bigelow provided this rather elegant description of a hurricane, published in the *Yearbook of the Department of Agriculture for 1898.*

The physical features of hurricanes are well understood. The approach of a hurricane is usually indicated by a long swell on the ocean, propagated to great distances and forewarning the observer by two or three days. A faint rise in the barometer occurred before the gradual fall, which becomes very pronounced at the center; fine wisps of cirrus clouds are seen, which surround the center to a distance of 200 miles: the air is calm and sultry, but this is gradually supplanted by a gentle breeze, and later the wind increases to a gale, the clouds become matted, the sea rough, rain falls, and the winds are gusty and dangerous as the vortex core comes on. Here is the indescribable tempest, dealing destruction, impressing the imagination with its wild exhibition of the forces of nature, the flashes of lightning, the torrents of rain, the cooler air, all the elements in an uproar, which indicate the close approach of the center. In the midst of this turmoil there is a sudden pause, the winds almost cease, the sky clears, the waves, however, rage in the great turbulence. This is the eye of the storm, the core of the vortex, and it is, perhaps, 20 miles in diameter, or one-thirtieth of the whole hurricane. The respite is brief and is soon followed by the abrupt renewal of the violent wind and rain, but now coming from the opposite direction, and the storm passes off with the features following each other in the reverse order. There is probably no feature of nature more interesting to study than a hurricane, though feelings of the observer may sometimes be diverted by thoughts of personal safety!

Hurricane of August 1, 1899, Carrabelle

After reaching the coast and maintaining very high velocities from the northeast backing to the North and West for 10 hours, the storm gradually abated leaving the town of Carrabelle a wreck.

The results to shipping were disastrous, 14 Barks (transport sail boats) and 40 vessels under 20 tons having been wrecked. The loss of life was amazingly small, the total being only six. The property loss, including vessels and cargo will amount to $500,000 (U.S. Weather Bureau, 1899).

The Second Thirty Years, 1901-1930[2]

This thirty-year period had less tropical storm and hurricane activity than the preceding (1871-1900) period or the following thirty-year period (1931-1960). From 1901 to 1930, there was a combined total of 39 tropical storms and hurricanes as compared to 63 during the previous 30 years (1871-1900) and 51 for the following thirty years (1931-1960). Storms during this period came primarily from the southwest.

U.S. Weather Bureau records (1901-1930) show that there were 22 hurricanes during this period; specific hurricanes are listed in Table 4. With the availability of more factual data published in the *Climatological Data* bulletin since 1897, information now becomes more accurate and detailed, consisting of actual reports for those years.

Looking at the first 10-year segment (1901-1910), the Great Hurricane of October 1910 did a loop north of the western tip of Cuba (Gray, 1949), passed through Key West and entered the coast near Ft Myers, where a low pressure of 28.20 inches of mercury was reported. This was probably one of the most destructive hurricanes to hit Florida.

At Key West, there was a 15 foot storm tide and Sand Key reported 125 mile per hour winds. The U.S. Army and Marine Hospital Docks were swept away at Key West in this hurricane, but little other narrative information is available on this storm except that

---

[2] Plates 4-6.

it made landfall near Cape Romano.

Actually, Key West, which is touted in stories and movies as a typical hurricane setting, is not all that hurricane-prone. The last encounter was 1987's Hurricane Floyd, about noon on the 12th of October. Highest winds were about 80 miles per hour and pressure was about 29.32 inches. Floyd's eye was reported at Key West, Marathon and Islamorada and was a weak category one storm. Before Floyd, it had been 21 years, all the way back to Hurricane Inez, since a hurricane had struck the Keys. In 1965, Hurricane Betsy sideswiped the "Conch Capital" as did Isbell, in 1964. These storms followed a 14-year lull during which the Keys were untouched by hurricanes. In 1950, Easy struck the Keys bringing to an end the area's 28 years of calm going back to 1919.

From 1871 to 1987, Key West was hit by 14 hurricanes or about 10% of the storms discussed here.

Other hurricanes during the 1901-1910 period were the hurricane of September 1906 which practically destroyed Pensacola and the hurricane of 1909. These two storms have good documentation which is worth further discussion.

Hurricane of September 19-29, 1906, Mobile-Pensacola Area

According to the 1906 U.S. Weather Bureau report this was a major storm.

> This was the most terrific storm in the history of Pensacola, or since the Village of Pensacola on Santa Rosa Island was swept away 170 years ago .... During the height of the storm, the water rose 8 $^{1}/_{2}$ feet above normal high water mark, being the highest known. The entire water front property was inundated; train service in and out of the city was completely paralysed ... Muskogee Wharf, belonging to the L&N Railroad Co., was broken in two in the middle, and the tracks on either side of the Main Creek were washed away [including thirty-eight coal cars] ...The greatest havoc was wrought along east Main Street, the south side of which has been completely washed away. The total damage from this hurricane will be three to four million dollars (equivalent to $80-100 million in 1990).

This hurricane made actual landfall in Alabama but affected Pensacola; because of this storm 164 people lost their lives.

Other storm notes by the U.S. Weather Bureau for the hurricane of September 1906 are from St. Andrews, Washington County.

> On the 26th, a tidal wave swept this place; the water was higher than any time during the past 19 years, and every wharf in St. Andrews was completely destroyed.

A report from Apalachicola, Franklin County:

> On the 27th, the wind blew a gale from the southeast, and on the 28th, it increased to a hurricane velocity. The amount of rainfall was 10.12 inches.

And from Galt, Santa Rosa County:

> The storm of the 26-28th was the worst ever known in this section; on the 26th, the tide rose 14 feet. Two lives were lost here.

Hurricane of October 6-13, 1909, Sand Key

Tannehill (1938) provides the following discussion of the October 1909 hurricane that struck Sand Key and resulted in 15 deaths:

> The hurricane of October 1909, was one of exceptional intensity. It recurved over the extreme southern tip of Florida, at which time it had attained tremendous force.
>
> The Weather Bureau had a station at Sand Key, Florida which was abandoned at 8:30 a.m., and supplies and instruments were carried to the lighthouse. The wind was then 75 miles an hour; shortly thereafter, the anemometer cups were carried away and the wind was estimated at 100 miles an hour. All the trees were blown down and at 9:35 a.m. heavy seas swept over the island. At 10:30 a.m., the Weather Bureau building went over and was swept out to sea. The lowest barometer reading was 28.36 inches. At Key West the barometer fell to 28.50 inches and the extreme wind velocity was 94 miles. Property damage there amounted to $1,000,000 [equivalent to $20 million in 1990]. About four

hundred buildings collapsed.

During the second 10-year segment (1911-1920), there were four hurricanes including one Great hurricane which deserves mentioning. Three of these, all with winds over 100 miles per hour, affected the Pensacola area again like the hurricane of September 1906.

The Hurricane of July 1916, Mobile-Pensacola Area

The U.S. Weather Bureau (July 1916) reported that:

> . . . . at 1 PM, a 92 mile per hour gale occurred with severe puffs from the southeast. The duration of the gale was extraordinary, and the total damage to the crops and the property will easily total $1,000,000 [equivalent to $20 million in 1990] for the section.

This hurricane made landfall in Mississippi where four lives were lost.

The Hurricane of October 1916, Pensacola

The barometric pressure in this storm was 28.76 inches at Pensacola.

> The wind instrument tower at the Weather Bureau Office blew down at 10:14 AM, after registering an extreme rate of 120 miles per hour at 10:13 AM. Oak trees that withstood the July storm were uprooted; about 200 trees throughout the city were blown down (U.S. Weather Bureau, October 1916).

The Hurricane of September 1917, Pensacola-Valparaiso Area

> This was a very severe storm, doing much damage on the coast and to crops. The lowest barometer reading, 28.51 inches, was a record for the Pensacola Station. The highest wind velocity during the storm was 103 miles an hour with an extreme rate of 125 miles

an hour from the southeast (U.S. Weather Bureau, September 1917).

The Great Hurricane of September, 1919, Key West

The following citation for the Great Hurricane of September 1919 was taken from NOAA (1987).

> The storm that passed over Key West on September 9 and 10 was, without question, the most violent of any recorded at this station. Property loss is estimated at 2 million (equivalent to 40 million dollars in 1990). In the terrific gusts that prevailed during the height of the storm, staunch brick structures had walls blown out, and large vessels which had been firmly secured, were torn from their moorings and blown on the banks (U.S. Weather Bureau, September 1919).

Lowest barometric pressure was 27.51 inches of mercury at Dry Tortugas with 300 lives lost in Key West where winds were 110 miles per hour. According to a recent NOAA report by Hebert, Jarrell, and Mayfield (1992) this storm ranked third among the most intense hurricanes to strike the United States this century until hurricane Andrew took over that ranking in August of 1992.

During the last ten years of the period from 1901-1930, there were six interesting hurricanes, including two Great Hurricanes which could be considered equivalent to category 4 hurricanes, according to the Saffir-Simpson Scale; some descriptions of these storms are briefly either quoted or described here.

The Hurricane of October 20, 1921, Tarpon Springs

> Great damage resulted at Tampa and adjacent sections from the combined effects of high winds and storm tides. The tide at Tampa was 10.5 feet, the highest since 1848. Eggmond and Sanibel Island were practically covered by water (U.S. Weather Bureau, October 1921).

Barometric pressure was 28.17 inches at Tarpon Springs and

winds were more than 100 miles per hour.

Only one hurricane and one tropical storm were recorded for Florida in 1925. The storm that came ashore near Tampa on November 30 was significant from a statistical standpoint—it was the latest any storm had hit the U.S. during hurricane season.

The Hurricane of July 26-28, 1926 Indian River

> The Center was near Palm Beach on the morning of the 27th, then north-northwestward. The high winds and seas sweeping before them boats, docks, boat houses and other marine property on the ocean front as well as that on the Indian River. Trees were uprooted, including citrus trees; houses were unroofed or otherwise damaged. The observer at Merritt Island remarks that there was a tremendous wave (this on the Indian River) and with the high wind all boats, docks, and other property from the river front were swept ashore ... (U.S. Weather Bureau, July 1926).

The Great Miami Hurricane of September 11-27, 1926[3]

> From the viewpoint of property loss, low barometric pressure, and maximum wind velocities at Miami, the hurricane of September, 1926, stands unchallenged in the meteorological records of the Weather Bureau, save only in respect to the loss of life at Galveston during the hurricane of 1900. The storm waters of the Atlantic united with the waters of Biscayne Bay and swept westward into the City of Miami......This was the most severe storm that ever visited this city. The extreme velocity was registered at 7:26 AM. The average velocity for the 20th was 76.2 miles an hour. Never before have hurricane winds been recorded for so long a time and never has the wind maintained a velocity of 100 miles for more than a hour (U.S. Weather Bureau, September 1926).

Winds and barometric pressure of this storm were 138 miles per hour and 27.61 inches of mercury, respectively.

The following excerpts are from copies of letters kindly donated

---

[3] See Figures 1-3.

by Mrs. Ruth Warner of Barefoot Bay, Florida, documenting her Grandmother's experience in the 1926 Miami Hurricane; her Grandmother was Mrs. Lucia Lawrence and the following are taken from her letters written in September and October 2, 1926.

The weather bureau broadcasted that a hurricane of great intensity was headed for the east coast, but that around Jupiter would be the center of the storm, but Miami got it.

About midnight, the wind was blowing a gale and the electric lights went out; everything in darkness.

When we got candles lighted, [we] found the water pouring through the ceilings on the rear half of the house so we knew the roofing was off.

With daylight Saturday came a lull in the wind for about 45 minutes. A good many [people] didn't know it was the center of the storm and so were fooled. Mrs. Moran (a friend at who's house they were staying) says the worst is yet to come but it will come from another direction. Sure enough the puffs soon began coming, but from the south east. Before that it was from the north east.

We all huddled in the dining room and kitchen until it was over, expecting every moment to feel and see the house going to pieces, at least, the front caving in as it rocked and swayed as the gusts struck.

We nailed the doors, watched as the screens and awnings go. Said we had done all we could do and left the rest with God.

The fury of the storm was terrible. It made such a peculiar muffled roaring sound in the air above. There are about 18,000 homes, either completely demolished or roofs torn off. About 5,000 injured and a good many more dead than the papers give, I believe. Probably a good many from the boats will never be found.

It's some mess to have all ones bedding blankets, clothing and bureau contents soaked at the same time.

The Hurricane of August 7-8, 1928, Indian River

Damage to property was heaviest from South Brevard to St. Lucie Counties... substantial houses were unroofed and frail ones were razed. Highways were flooded and badly washed. Many

> bridges were undermined requiring replacement. Many citrus trees were uprooted, the loss of fruit estimated at 1,000,000 boxes. Large oaks, sentinels of a century, were uprooted (U.S. Weather Bureau, 1928).

The Deadly Great Lake Okeechobee Hurricane of September 6-20, 1928[4]

This category 4 storm tracked across Lake Okeechobee's northern shore, causing the shallow waters to reach heights of more than 15 feet. This surge was forced southward, causing terrible flooding in the lowlands at the lake's south end. This area was farmed primarily by migrant workers. Thousands of migrant farmers died as water rushed over the area. After the storm, the Red Cross counted 1,836 dead, but still more bodies and skeletons were discovered in later years. The barometric pressure was measured at 27.43 inches. To prevent future similar disasters, dikes were built around the lake by the U.S. Army Corps of Engineers. The 1928 storm caused $25 million (equivalent to $300 million in 1990 dollars) in damage.

From the Hebert et al. (1992) report this hurricane ranked second among the deadliest hurricanes to strike the U.S., and was ranked fourth among the most intense hurricanes to strike the U.S. But this storm falls to fifth place, as far as intensity, after Hurricane Andrew, which struck south Florida in August 1992 with a low barometric pressure of 27.23 inches.

The Hurricane of September 28, 1929, Key Largo

Tannehill (1938) provides the following account of this hurricane striking the Keys.

> The center passed over Key Largo on the 28th, barometer about 28 inches and wind estimated at 150 miles an hour. There was a ten-minute lull as the center passed. At Long Key the barometer was 28.18 inches. At the Everglades, the wind was esti-

---

[4] See Figures 4-5.

mated at 100 miles an hour, barometer 28.95 inches. The storm reached Panama City on the 30th, barometer 28.80 inches.

Although there was enormous damage at Nassau in the Bahamas and many lives were lost there, its course in Florida was such that damage probably did not exceed $500,000 ($6 million in 1990 dollars) and only three lives were lost. The population had been thoroughly warned by the Weather Bureau and there had been ample time for all possible precautions.

The Third Thirty Years, 1931-1960[5]

This thirty-year period had more storm activity than the previous thirty-year period (i.e., 51 total storms compared to 39). There were 21 hurricanes, almost the same as reported for the previous period. However, tropical storms numbered 30 as compared to 17 for the previous thirty years, which accounts for the high number of total storms.

The temporal distribution of hurricanes from 1931-1960 is interesting. While there were few hurricanes from 1931-40 (six) and 1951-1960 (three), there were 12 hurricanes for the 10-year period 1941-1950 alone. This made the 1941-1950 segment the most destructive and costliest period to that date in terms of equivalent dollar value since records were kept for the state. Yet ten years later, in 1960, one single hurricane, Donna, a Category 4, was even more costly and destructive than all the storms occurring in the total 10 year period from 1941-1950 (Hebert et al., 1992).

Looking at the first 10 years (1931-1940), out of a total of 6 hurricanes, there were two Category 3 storms and one Category 5 hurricane in 1935 which was one of the only two Category 5 hurricanes to ever hit the U.S. coast with that intensity - the other was Hurricane Camille which struck Mississippi in 1969. Hurricane Allen, which struck Texas in 1980, reached Category 5 intensity three times during its path but weakened to Category 3 at landfall (Hebert et al., 1992).

---

[5] See Plates 7-9.

The Major Hurricane of September 1933, Jupiter

In July and September 1933, two hurricanes entered the east coast of Florida within a short distance of each other. The second of these two, which occurred on Labor Day, deserves review.

> There was much property damage on the east coast from Vero Beach to Palm Beach; a few houses were totally demolished, quite a number blown off their blocks. More than the equivalent of 4 million boxes of citrus were blown from the trees statewide. The property loss in Indian River, St. Lucie, and Palm Beach Counties probably was about 2 million dollars ($25 million in 1990) (U.S. Weather Bureau, September 1933).

In addition to the above report, an elderly citizen from Ft. Pierce recalls that the 1933 storm was the most devastating in the history of Ft. Pierce (Yanaros, 1986).

In 1935, two hurricanes visited southern Florida. The first was the Great Labor Day Hurricane of September 1935 and the other was the October 30th through November 8th, storm called the Yankee Hurricane because it came in from a northeasterly course and struck the extreme south Florida coast and the west coast.

The Great Labor Day Hurricane was the most violent in the history of Florida and the United States. It was the only Category 5 storm ever to strike Florida; its central barometric pressure of 26.35 inches of mercury was the lowest ever recorded at that time in the western hemisphere. (As of 1988, Hurricane Gilbert, which did not affect Florida, has the record 26.22 inches of mercury for the lowest barometric pressure in the western hemisphere).

The following excerpts from the 1935 storm are quoted from Mr. Gray's 1949 paper entitled *Florida Hurricanes*.

> No anemometer reading of the wind was obtained, but the gradient formula gives 200-250 miles per hour and the engineer's estimate by stress formula is in substantial agreement ... the path of destruction was less than 40 miles in width. More than 400 people were killed, most by drowning. The tracks of the Flagler Railroad were washed from the Long Key viaduct at an elevation of 30 feet above mean low water. A survey by the U.S. Engineers some time after the storm indicated that the tide level never reached the rails

> there, but the hurricane surge superimposed on the tide probably assisted in carrying the tracks away.

(Maximum storm surge with Hurricane Camille was 24.2 feet).

In 1938 Tannehill described a tragic event of the storm.

> A rescue train that was sent to remove World War I veterans and residents from the Florida Keys, on September 2, 1935, was swept from the tracks by the hurricane and the storm wave. [6]

The following 10-Year period, 1941-1950, was the most devastating in Florida's history since records were kept. Out of 12 hurricanes, 11 of these took place between 1944 and 1950. In this relatively short period there was one category 4 in 1947, and six category 3 hurricanes, one each in 1944, 1945, 1948, 1949, and two in 1950; a Category 1 storm struck Ft. Myers on the west coast in 1946. All are discussed here.

The following is a quotation describing the 1944 storm, taken from U.S. Weather Bureau in their report of October 1944.

> Dangerous winds extended fully 200 miles to the right or east of the center, about 100 miles to the left or west, thus affecting the entire peninsula of Florida. [Even at Dry Tortugas, barometric pressure was 28.02 inches of mercury.] Winds of hurricane force velocity surrounded the central core, with gusts up to 100 mph at Tampa and Orlando. Tides were high from Sarasota southward on the Gulf and from Melbourne northward on the Atlantic, Naples, and Jacksonville Beach both reported 12 foot tides. Citrus loss was over 21 million boxes (average harvest was 80 million). Throughout the state there was damage to telephone, telegraph and power lines, trees, roofs, chimneys, signs, and radio towers. Of the interior cities, Orlando seems to have suffered the most damage, being estimated at over one million dollars.

The next hurricane of importance entered the coast in September 1945 at Homestead, curving northward right up through the center of

---

[6] See Figures 6 and 7.

Florida. During the course of the storm, it decreased in windspeed, but maintained itself as one with minimal hurricane force. It also remained over land to exit near Jacksonville Beach. The tragic event with this storm was the destruction at Richmond, Florida, of the three Navy blimp hangars which were used as evacuation shelters for 25 Navy blimps, 183 military planes, 153 civilian planes and 150 automobiles. The three great hangars were torn to pieces at the height of the storm, and then caught fire and burned with all their contents; the total loss was estimated at 35 million dollars (U.S. Weather Bureau, September 1945).

Then came the Category 4 hurricane of September 17, 1947, clocking the highest recorded windspeed, except for Hurricane Andrew in 1992, in Florida's history with a 1-minute maximum windspeed of 155 miles per hour, recorded from a reliable instrument at the Hillsboro (Pompano Beach) light station. The following describes this exceptionally strong hurricane (U.S. Weather Bureau, September 1947).

> Hurricane force winds were experienced along the Florida East Coast from about Cape Canaveral to Carysfort Reef Light (south of Miami), a distance of about 240 miles, while winds of 100 miles per hour, or over, were felt from the northern portion of Miami to well north of Palm Beach, or about 70 miles. This classifies this hurricane as one of the great storms of recent years.

This September 1947 storm had a barometric pressure of 27.97 inches of mercury at Hillsboro, with tides at Clewiston and Moore Haven of 21.6 feet and 20.9 feet, respectively. This storm was nearly as bad as the 1928 hurricane at the lake. Fifty-one people died.

During October 9-16, a hurricane came across western Cuba into southwest Florida, northeast into the Atlantic around Palm Beach. It was a Category 1 and was seeded for the first time. It split in two in the Atlantic and the worst part hit Savannah, Georgia.

Two hurricanes occurred in 1948. The first one ran from 18 September to 25 September and was classified as a Category 3. The system started just west of Jamaica and moved west to northwest then north over western Cuba into the Florida Straits. It struck Florida near Everglades City in the 10,000 islands, then moved northeast through Florida to emerge into the Atlantic near Jupiter. A tornado was reported in Homestead on the 21st of September. Lowest barometric pressure was 28.44 inches, and top winds were 122 miles per hour. The hurricane killed 3 people and caused 105 million dollars damage.

The second 1948 storm ran from the 3rd to the 15th of October and started just off the Nicaraguan/Honduras coast in the northwest Caribbean Sea. This hurricane also moved across western Cuba into the Florida Straits and even crossed the September hurricane's path near the coordinates 24.0N and 82.0W. This storm passed through the Keys and extreme south Florida into Grand Bahama Island. At about 31N latitude it did a gigantic loop in the middle of the Atlantic and finally became a non-tropical cyclone. A tornado was reported in Fort Lauderdale on the 5th of October. Lowest barometric pressure was 28.92 inches, and top winds were around 90 miles per hour.

In August 1949, another major hurricane, taking a course similar to the Great Hurricane of September 1928 entered the coast near Palm Beach. It was the worst hurricane felt in the Lake Okeechobee area since 1928. Hurricane force winds were reported at St. Augustine, Cape Canaveral, and Melbourne, and winds of 120 miles per hour or greater were felt from Stuart to Pompano. The highest recorded wind speed gust, 153 miles per hour, was at Jupiter, only 2 miles per hour less than the wind speed record set on September 27, 1947. The amount of damage in dollars, 45 million (equivalent to $270 million in 1990), was almost twice that of the 1928 hurricane. The storm was not classified as being among the Great Hurricanes in Florida's history, but it fell into the category of only being slightly below them (U.S. Weather Bureau, August 1949). Tides were 24 feet and 23 feet at Belle Glade and Okeechobee, respectively.

Finally, here are some brief quotations, taken from the U.S. Weather Bureau reports, 1950, for the last two major hurricanes of the 1941-1950 decade; during this period the storms were named using World War II phonetic alphabet: Able, Baker, Charlie, Dog, Easy, etc.

Hurricane Easy, September 1950, Cedar Key

> Old residents say this was the worst hurricane in 70 years... half of the houses were severely damaged or destroyed... The fish-

> ing fleet upon which the town depends for a livelihood, was completely destroyed by wind and waves... The tide in Tampa Bay rose 6.5 feet., the highest since 1921.

This hurricane looped twice on the west coast, had top winds of 125 miles per hour and a barometric pressure of 28.30 inches, caused 38.7 inches of rain at Yankeetown in the September 5-6 period, and brought unfounded accusations of seeding by the Weather Bureau from residents of the area.

Hurricane King, October 1950, Miami

> The path of principal destruction was only 7-10 miles wide through the greater Miami area and northward to West Fort Lauderdale on the 17 of October. It was at first reported that the damage was the result of a tornado or tornadoes.... after careful inspection there was no evidence of tornado action ... It was simply that of a small violent hurricane.

Gusts were 150 miles per hour at Miami and 138 miles per hour at Ft. Lauderdale. Barometric pressure fell to 28.20 inches and tides were 19.3 feet at Clewiston. Three people died during this storm.

Following is a personal eyewitness report on Hurricane King described by one of the authors (J.M. Williams).

> To begin, this storm, Hurricane King, formed down in the northwest Caribbean not too far from Swan Island. It curved its way northward to clip the west tip of Jamaica. From there King traveled almost due north across Cuba to slam into Florida close to Miami and made passage through western Ft Lauderdale. I was home on leave from Army duty on the 17th of October, 1950, visiting my folks who lived in Country Club Estates, which is now Plantation. This was my first hurricane on land—I had been in one on a ship at sea coming back from Occupation Duty in Europe. In the afternoon of the 17th, rain was coming intermittently in sheets and the wind was gusting pretty high. Then, as if nothing was going on, it would calm down and the sun would come out. We were from Iowa where when it looked stormy, you were going to get it! That, I found out didn't mean anything down here in Florida. My mother had two cats who were progressively getting noisy

and mean. We found out later that they were affected by the barometric pressure drop as the storm approached. By the time the storm hit they were climbing the walls! The house was CBS block construction so we felt okay because shutters had been installed some years ago. On the front porch were aluminum jalousies.

The main part of King hit us in the late evening and it was really something, to me at least! The street in front of the house was gravel back then and the winds picked the rock up and blasted the front of the house! The noise on those aluminum jalousies was so bad we couldn't hear each other talk. There was a lot of lightening in the storm and we could see out through the shutters. Newspapers were flying all over the place. The only trouble however, the newspapers were not newspapers but were tiles off the roof.

We went outside during the eye and I found out what everybody had always said about the eye: we could see stars, the moon, and a few clouds, and we could feel a slight breeze. We detected a smell, some said was ozone. But ozone is odorless! Nevertheless we experienced the Hurricane Eye Smell.

The famous backside of the hurricane came right away, like now, and we rode out the remainder of the storm like we did the first part. Now the wind came from the other direction and it loosened up everything.

After the passage of Hurricane King we took a long look at the damage. Out of six fruit trees, only one was still standing. About half of the roof tiles were lost and would need to be replaced. The aluminum jalousies were dented and stripped of all paint. Debris could be seen all over the place. As the area was wooded in that period of time, trees were down here and there. A big tree, about two feet in diameter took down the power and telephone lines. I had returned to Fort Benning by the time power had been restored. We heard that a tornado was running around in the eye of King and wiped out a trailer park in the town of Dania, south of us. Some people were killed due to King.

The last 10-year period, 1951-1960, of the 1931-1960 thirty-year segment was marked by a sharp reduction in major hurricane activity; it was during this period when the Weather Bureau began giving hurricanes female names in 1953.

In October 1951, Hurricane How, as a tropical storm, crossed mid-Florida. In 1952, a tropical storm crossed south Florida in February. In 1953, tropical storm Alice struck northwest Florida in

June. Another tropical storm crossed south Florida in August. Yet another tropical storm crossed north Florida in September and Hurricane Florence hit northwest Florida also in September. Tropical storm Hazel crossed mid-Florida in October to finish out 1953.

In 1956, Hurricane Flossy struck northwest Florida in September. In 1957 two tropical storms hit the same region, one was unnamed and one was named Debbie. In October 1959 two more tropical storms came ashore in Florida—Irene into northwest Florida, and Judith crossing mid-Florida. None of these were of major consequence.

Hurricane Donna stole the show in 1960, while Brenda, as barely a tropical storm, crossed north Florida in September, and Florence also a very weak storm, crossed south and central Florida a week earlier.

Hurricane Donna ranked fifth, prior to Hurricane Andrew in 1992, among the most intense hurricanes ever to strike the U.S. this century (Hebert et al., 1992). Except for the western Panhandle, where Flossy with a barometric pressure of 28.93 inches affected Pensacola in 1956 with gusts at 98 miles per hour, Donna was the first hurricane to have a major affect on Florida since Hurricane King in 1950.

Hurricane Donna caused $300 million ($1.9 billion, 1990 dollars) in damages to the state and was one of the most destructive hurricanes to affect Florida in modern times (Dunn and Miller, 1964; Hebert et al., 1992), although Hurricane Andrew in 1992 will replace Donna as Florida's most damaging storm.

At Conch Key, pressure was 27.46 inches on the 10th of September, 1960, and tides were 13 feet 40 miles northeast and 20 miles southwest. Donna was at her peak here, moving only 8 miles per hour. The storm killed three people in the Keys. Top winds of 180-200 miles per hour were recorded in the Keys, with gusts to 150 miles per hour at Everglades City and Naples. In central Florida, the pressure was 28.60 inches at Lakeland, 28.66 inches at Orlando, and 28.73 inches at Daytona. Barometric pressure was 28.05 inches at Ft. Myers. Fifty people died when a U.S. airliner crashed off Dakar, Africa, at the beginning of the storm.

Following are some quotations from the U.S. Weather Bureau records, (September 1960) about this hurricane.

Storm damages range from very severe in the Middle Keys and the southwest coast from Everglades City to Punta Gorda, to relatively minor in northwest Florida and points north of the storm track. At Naples tides pushed inland to the center of the city damaging buildings and smashing docks all along the intrusion. Everglades City, a town that had been largely evacuated was also inundated by storm tides and about 50% of the buildings in that city were destroyed by tides and winds. Even well outside these areas, the wind toppled thousands of trees, demolished many weaker buildings, blew off or damaged roofs, and shattered many windows. Power and communication facilities fell throughout central and south Florida. Grapefruit losses were between 25 and 35% of the state's crop. Gusts of 99 miles per hour recorded at the FAA tower in Daytona Beach marked Donna's exit from Florida, having retained hurricane status throughout its entire passage in Florida.[7]

Donna inflicted major ecological damage. Dunn and Miller in 1964 reported that one of the world's largest stand of mangrove trees was 50% wiped out in many areas and that 35 to 40% of the white heron population was killed.

In Everglades National Park, a monument on the road to Flamingo reminds visitors today about Hurricane Donna.

The Last Thirty Two Years, 1961-1993[8]

There were 27 storms during this 32-year segment. Comparing these figures with the 21 hurricanes and 30 tropical storms for the previous 30 years, one can easily see the overall reduction in both hurricanes and tropical storms (Table 4). In the 1961-1992 period, 6 hurricanes (category 3 or higher), Betsy, Eloise, Elena, David, Inez, and Andrew, occurred as compared to 11 from 1931-1960. Hurricanes Inez and Kate did strike Florida but were categories 1 and 2 storms then. Inez and David were Category 4 storms in the Caribbean. Hurricanes Juan and Elena, in 1985, affected northwest Florida without a strike.

During the first 10-year period, 1961-1970, seven hurri-

---

[7] See Figures 8 and 9.

[8] See Plates 10-13.

canes—Cleo, Dora, Isbell, Betsy, Inez, Alma, and Gladys—hit Florida, a sharp increase over the previous 10-year period, 1951-1960. While there were no storms from 1961-1963, 3 hurricanes struck Florida in 1964 alone, making this year the costliest ($350 million and more, which is equivalent to $1.75 billion dollars in 1990) so far in Florida's history.

In 1962, Alma as a tropical depression passed Florida's east coast on 26 August. In 1965, a tropical storm crossed northwest Florida on 15 June from the Pacific.

In late August 1964, Hurricane Cleo was the first hurricane to strike the Miami area since Hurricane King in 1950. Cleo moved up the peninsula about 20 miles inland paralleling the east coast. It produced 138 mile per hour gusts at Bahia Mar Marina, Ft. Lauderdale, and knee-deep water was observed in some locations. Due to its small size, Cleo soon weakened to below hurricane strength around the Fellsmere-Melbourne area, yet the total storm damage was estimated at $125 million ($600 million in 1990 dollars). Cleo sailed through Georgia, South Carolina, and North Carolina to break into the Atlantic on the 1st of September. She regained hurricane status on the 2nd, but died in the north Atlantic near Nova Scotia on the 5th of September.

According to the U.S. Weather Bureau report of August 1964, principal losses caused by Hurricane Cleo were from glass and water damage in the Miami Beach area, and agricultural losses in the Indian River citrus belt.

Author John M. Williams provides the following personal account of Cleo whose path was tracked semi-hourly through southeast Florida (Figure 10), in the Ft. Lauderdale area.

> Cleo was the worst in the southeast coastal area. Cleo was of Cape Verde vintage and traveled through the Atlantic and the Caribbean as a "textbook" storm. Between Jamaica and Haiti, however, she turned northward into the Guantanamo Naval Base in Cuba causing considerable damage there and in Cuba. Passing across Cuba brought the usual decrease in strength, but once into the Florida Straits she regained her hurricane status. The following is my eye-witness report on the passage of Cleo in western Ft. Lauderdale area.
>
> There were winds and rain all day of the 26th August. Some gusts were in the 60 miles per hour category. It seems that I had

the only ladder in the neighborhood and since the people there knew I was in the weather business, a line of them formed as I was finishing the preparations on my house. I didn't see the ladder again until after the storm had passed, but I heard it got as far as two blocks away.

I knew the storm was going to hit this area after dark so we decided to have supper and get all the other amenities out of the way. We put all the kids to bed early but that didn't last for long, after all, it was their first hurricane! It was lucky that we toweled up all the doors and had the shutters on the windows because at the height of the storm we had water coming in the front door and through some of the windows—we had glass jalousies throughout the house!

It peaked late in the evening just before the eye passage with gusts to 130 miles per hour at my location and there was considerable lightning, along with that tremendous roar. You could see almost like daylight through the shutters.

My children will never forget the 'little leaf', obviously sheltered by the house, hopping across the yard, in the opposite direction from the wind.

About ten minutes before the eye, a Florida room aluminum shutter, about 3 by 8 feet, ripped off the house next to mine. It slammed into the corner of my house and ricocheted out into my front yard. I had a small palm tree out there which was bending over from the winds and the shutter managed to wedge itself between the tree and the ground.

The eye passage lasted one hour and twelve minutes at my location.

I opened the door at that time to a rush of water about two inches deep. While the wife mopped that up, I stepped off the porch into nearly knee-deep water and waded to the palm tree. As hard as I tried, I couldn't free the shutter from the tree.

I could see the stars in a beautiful sky about me and there was that unmistakable stillness and smell that only happens in the eye of a hurricane!

The guy across the street yelled over to me that he had lost all the glass jalousies from his Florida room and had to move inside the house. He had only taped his windows!

I tugged again and again at the tree and big shutter but to no avail. I couldn't move it. I checked around the house and everything seemed all right or passable. But now it was time to get back in the house because the backside of a hurricane comes on like

'Gang Busters'!

Since the wind comes from the opposite direction and right now, it is there before you know it! And it came! The palm tree straightened up and the big shutter came loose and was last seen, in lightning flashes, heading north, up over the house across the street! We never saw it again. The back side of the storm was drier than the front but not by much.

Since the house leaked (all houses leak in a storm like Cleo), we had a lot of mopping to do. The pea-rock on the flat roof of the garage was all gone and there was a dent in the decklid of the car parked in the carport; something was flying around loose. When the water subsided, it left a mass of debris all over the place and power was off in some parts of town for five days. Our power came on again by late afternoon of the 27th of August one day after the storm passed by. There was widespread damage throughout the area but only occasional catastrophic type.

In a few days, we had the place almost cleaned up; I had my ladder back and the kids still wanted to know what happened to the 'little leaf'. For a period after the storm when I mowed the lawn, the clippings were a combination of grass and pea-rock shrapnel.

In September 1964, just a few weeks after Hurricane Cleo, Hurricane Dora struck the Florida coast at a near 90 degree angle from the east at St Augustine, Florida. It was the first hurricane to do so, north of Stuart, since the Great Hurricane of 1880. The hurricane's winds of 125 miles per hour at St. Augustine resulted in a 12-foot storm tide which swept across Anastasia Island (St. Augustine) and also produced a 10 foot storm tide at Fernandina Beach, and Jacksonville. These massive storm tides caused extensive beach erosion, inundated most beach communities, washed out beach roads, and swept buildings into the sea. There was also considerable flooding along the St. Johns River in Jacksonville. Total damage was estimated at $250 million dollars (more than $1 billion in 1990 dollars) (U.S. Weather Bureau, September 1964).

Hurricane Isbell, while not a strong hurricane, struck Florida in October 1964 and is described as an eyewitness account by one of the authors, J.M. Williams.

This storm grew down south of the western tip of Cuba and proceeded northeast, across Cuba, the Florida Straits, and into the Ten Thousand Islands region of southwest Florida. From there,

Isbell took a more northeast course across Florida. On the evening of October 14th, the storm passed just northwest of Fort Lauderdale, Florida. Winds were 50 to 60 miles per hour with a recorded gust of 120 miles per hour. Many tornadoes, spawned by Isbell, caused as much damage as the hurricane did. Rains were extremely heavy in the early period of the storm but slacked off to nearly dry conditions at the end. A tornado, less than a block from where I lived, tore the whole Florida room, constructed of block, off a house. Isbell passed out to sea between the cities of Palm Beach and Vero Beach and dissipated in the Atlantic.

The following year, in September 1965, Hurricane Betsy, a Category 3 storm, struck extreme southern Florida from the east. Wind gusts up to 60 miles per hour were reported as far north as Melbourne. In south Florida, an observer at Grassy Key reported winds of 160 miles per hour before the anemometer was blown away at 7:15 AM on the 8th September. Six to eight foot storm tides and wave action caused considerable flooding between greater Miami and the Palm Beaches; rising waters flooded extensive sections of Key Biscayne, covering virtually all of the island (U.S. Weather Bureau, September 1965). (See Photographs).

Hurricane Betsy, (Figures 11 and 12) was unique and formed far out in the Atlantic around the 27th of August, and was obviously a Cape Verde type hurricane. After moving west for a few days, it developed an erratic course starting around Puerto Rico (Sugg, 1966). The path was a zig-zag, generally in a northwest direction to a point about 300 miles almost due east of Cape Kennedy, (as the Cape was known in those days). She became stationary there for nearly two days, then suddenly moved in a south-southwest direction which took her right into the central Bahamas. Just east of Nassau, Betsy stalled again. For 20 hours, winds of 120-140 miles per hour buffeted the area causing death and destruction.

The following eye-witness (J.M. Williams) report is about the passage of Betsy in western Ft. Lauderdale.

> During the 7th of September we were intermittently pelted with rain and strong wind gusts. Nassau is only about 150 nautical miles from Ft. Lauderdale and since Betsy was a large 'Cane', we were getting all sorts of weather in the area.

During the early morning of the 8th, we were getting rain in sheets with several gusts in the 125 miles per hour category. Sustained winds easily hung in there between 65 or 90!

Even though we did not experience the eye, things would calm down to almost sunshine conditions—but this would not last long.

There was a lot of flooding and house seepage but not as bad as last year's Cleo. Betsy's eye, which was huge and about 40 miles in diameter, was south of us and our pressure bottomed out at 29.12 inches.

This combination of pelting rain and heavy winds continued all day long, and even at supper time it was still not advisable to venture outside. Our power was off for more than ten hours and the usual mass of debris was all over the place. There was a lot of orange and grapefruit damage as well as damage to other crops; again most of the pea-rock was blown off the garage roof.

Our place was wet for a long time and I recorded more than eight inches of rain for the passage period. When there is no break up to the continuity of a storm (the eye), you get the effects all the time: more rains, more winds, more everything.

Inez was a Cape Verde type hurricane with a classic track through the Caribbean, across Haiti and Cuba and into the Florida Straits. From there she earned the name, "the Crazy One" by the National Hurricane Center. She took a very erratic course, first north, then south, then east, and finally west and this had everybody's fingernails completely gone!

Before she died in the mountains near Tampico, Mexico, Inez had killed more than 1500 people, had recorded top winds of 190 miles per hour, and planted a barometric pressure of 27.38 inches (from air reconn) in the books! Back then, that was called a "Severe" Hurricane. Today that would be a strong category 4.

The following eye witness report (J.M. Williams) is of the passage of the storm in western Fort Lauderdale, Florida.

I had put the wife and kids to bed early that night and told them that Inez was heading northeast. As erratic as it had been though, I was going to stay up and keep a check on it. I was off duty so there was nothing else to do, and I was a Storm-Hunter anyway. I was glued to the weather radio, TV, barometer and the rest of the instruments at my station. But as enthusiastic as I was about the whole thing, I was guilty of dozing off two or three times.

The winds here were gusting more than 40 miles per hour and I had pulled down the shutters just in case.

At 0800, 3 October, the pressure at the house was 29.65 inches, temperature was 78°F, Dew Point was 78°F, humidity was 100%, winds were north at 29 miles per hour sustained and it was overcast with rain. Inez was 93 miles east-northeast of Miami moving north-northeast at 7 miles per hour. We had it made.

At 1100, the storm was 75 miles west-northwest of Nassau moving north-north east.

At 1400, she was stationary about 85 miles west-northwest of Nassau.

At 2300, Inez was drifting slowly south-southwest pushing 25 foot seas as reported by a Coast Guard Cutter. The Southeast Florida coast had gusts of more that 55 miles per hour. I dozed a couple of times even though I knew it was now coming this way.

At 0345, 4 October, I awoke to shutters rattling and pelting rain! Winds were gusting more than 60 miles per hour. Barometer was 29.59 inches, temperature was 75°F, dew point was 73°F, humidity was 91% and it was overcast with thunder and lightning!

At 0700, Inez was 45 miles southeast of Miami with winds of at least 85 miles per hour and moving west at 7 miles per hour.

At 1100, the storm was moving west-southwest at 8 miles per hour with gale-force winds 175 miles to the north, and 100 miles south. Here, we had hurricane gusts frequently and gales with heavy rains all day. U.S. Highway No. 1 in the Keys was under water. The eye of the storm was 30 miles in diameter.

At 2000, my barometer was reading 29.67 inches and I was still getting gusts in excess of 45 miles per hour. By 1500, 5 October, Inez was stationary near Dry Tortugas with winds of 120 miles per hour. From there, she finally continued west to Mexico.

We got a bit of minor damage on the house and there was a lot trash to pick up around the yard. Everything was wet for a few days, however, we considered ourselves lucky!

The earliest hurricane to hit the U.S. was Alma. She struck northwest Florida June 9, 1966.

During the 18th and 19th of October, 1968, Hurricane Gladys struck the west coast of Florida between Bayport and Crystal River about midnight on the 18th.

Gladys formed in the western Caribbean near Swan Island and

steadily move in a north track across western Cuba, over Dry Tortugas and into Florida's west coast.

Dry Tortugas and Plantation Key both reported winds near 90 miles per hour. The storm's forward speed was about 15 miles per hour. Tides along the west coast were 6.5 feet above normal causing beach erosion and flooding mostly between Clearwater and Bayport.

Maximum gusts were over 100 miles per hour and lowest pressure was 28.76 inches. Citrus was heavily damaged and mobile-home damage was extensive, as usual, as far inland as Ocala. Gladys broke out into the Atlantic near St. Augustine having killed 3 people in Florida and one in Cuba. One more death was added in Nova Scotia and the total damage was nearly $17 million in 1968.

The next to the last 10-year period, 1971-1980, had the lowest storm total of the 122-year history. Three hurricanes and one tropical storm. The three Florida hurricanes were Agnes, Eloise and David.

Hurricane Agnes, which occurred in 1972, was barely a Category 1 hurricane in Florida but resulted in major devastation in the middle, southern, and northeastern states, and caused 122 deaths and six billion dollars damage in 1990 dollars. Agnes struck the Florida panhandle, then merged with another system in the mid-U.S., triggering torrential rains and extreme flooding throughout the entire eastern seaboard.

The threat of a hurricane usually diminishes rapidly as it moves inland and loses its oceanic heat source, however, sometimes the storm will encounter an environment that supplies an auxiliary source of energy to maintain strength far inland. Such is the case with Agnes, which from landfall near Apalachicola, Florida, where losses were less than $10 million, she traveled nearly a thousand more miles to become one of the most destructive storms in U.S. history.

Hurricane Eloise, in 1975, was a Category 3 hurricane. Hurricane David, in 1979, had weakened from a Category 4, to a Category 1 hurricane when it struck Florida, but David still caused over $400 million in damage.

Hurricane Eloise, which came in September 1975, made landfall about midway between Ft. Walton Beach and Panama City (Balsillie, 1985). It was the first direct hit by a major hurricane in the 20th Century in that area.. Measurements of high water marks by the U.S. Army Corps of Engineers indicated hurricane tides of 12-16 feet above mean sea level. Eglin Air Force Base, 20 miles west of the center, reported the highest sustained wind of 81 miles per hour when the instrument failed; 14.9 inches of rain fell. However, maximum sustained winds were estimated at about 125 miles per hour with gusts to 156 miles per hour. The combined effects of winds and tides undermined or demolished numerous structures along the beach from Ft. Walton Beach to Panama City (North Atlantic Tropical Series, volume 26, 1975); the lowest barometric pressure was 28.20 inches. Damage was over $1 billion in 1990 dollars from this category 3 hurricane; there were 21 deaths in the United States.

In 1979, the National Hurricane Center decided to integrate Male and Female names for the hurricanes in the Atlantic for the first time.. Bob was the first Atlantic hurricane and the 'guys' decided to out-do the 'gals'....and they did!

In September 1979, Hurricane David moved inland south of Melbourne on the east coast and then northward along the Indian River to exit at New Smyrna Beach. It was the first hurricane to strike the Cape Canaveral area since the hurricane of 1926. Severe beach erosion from a near five foot storm tide was reported in Brevard County and the southern portion of Volusia County. Some homes, businesses, and public buildings were severely damaged or destroyed, however, most of the damage, though widespread, was minor because the strongest winds were just offshore over the adjacent Atlantic Ocean.

Figure 13 shows Hurricane David from the 22,000 mile high Geosynhcronus Operational Environmental Satellite (GOES) orbiting the earth. At this point, August 31, David was about to make landfall on Hispaniola about 1800 EST. Winds were near 150 miles per hour and the central pressure was 27.34 inches. Earlier, about 125 miles south of Puerto Rico, sustained winds of 150 miles per hour and central pressure of 27.29 inches was David's strongest point, then rated as a category 4 hurricane.

On 1 September at 0600, David broke into the sea north of Hispaniola and Haiti. Winds were down to about 75 miles per hour after crossing a 10,000-foot mountain in the Dominican Republic. Later that day, a hurricane watch was posted for south Florida with the weakened storm some 350 miles southeast of Miami. In the late evening of the same day, hurricane warnings were up as the now strengthened hurricane, with 90 mile per hour winds, was 300 miles from Miami.

At 0700 on September 3rd, David was 35 miles east of Ft. Lauderdale with 85 mile per hour winds and a pressure of 28.85 inches. Ft. Lauderdale experienced torrential rain, in squalls with gusts over 75 miles per hour. Since the eye and strong side of the storm were over the ocean, this condition kept up most of the day.

Figure 14 shows David, at about 1800 that evening. David made landfall about 20 miles south of Melbourne with 90 mile-per hour winds and central pressure of 28.75 inches, a Category 1 hurricane.

From there, the hurricane made it to Savannah, Georgia, before downgrading to a tropical storm, on September 4th. On the 7th, David was no longer a threat and died near Newfoundland (see Figure 14a).

Fatalities were: 5, United States; 7 Puerto Rico; 56 Dominica; and 1200 in the Dominican Republic. The damage was $5 million (1990 dollar value) in the U.S. and Florida.

While Hurricane Frederic (1979) did not strike Florida directly, hurricane warnings extended over to Panama City on September 11th, and gale warnings were displayed south to Cedar Key (Balsillie, 1985).

Frederic had a development that was similar to David. This caused much apprehension because people were not ready for another storm so soon, one week, after David.

The final thirteen years (1981-1993) of the thirty-two year period, had several storms and hurricanes, marking an upswing in overall storm activity (Table 4). Some of the storms became hurricanes after they passed Florida, and will be mentioned briefly here.

On August 17 and 18, 1981, tropical storm Dennis struck Florida. Dennis started as a tropical storm southwest of the Cape Verde Islands on the 6th of August, and continued at this level near Barbados. Dennis became a tropical depression south of Puerto Rico and then turned into a mere disturbance. Just west of Jamaica, Dennis regained tropical storm status, turned north and slammed into the southwest Florida coast. The track was up through central Florida to become stationary between Ft. Myers and southwest Lake Okeechobee. Southeast Florida had 10 inches of rain, and Homestead had 20 inches. Winds were more than 55 miles per hour. Finally, Dennis moved across the lake and out to sea near Melbourne and Cape Canaveral, Florida.

On the 20th, east of Cape Hatteras, Dennis became a hurricane.

On August 25th, 1983, tropical storm Barry struck Florida. This storm crossed Florida on a track from Melbourne to Tampa on the 25th, first as a tropical storm and then as a tropical depression. After crossing the Gulf, Barry became a hurricane southeast of Brownsville, Texas, on the 28th of August.

Hurricane Diana, which gained hurricane status on September 10, 1984, scraped the Florida coast between Daytona and Jacksonville on the 9th-10th as a tropical storm; winds were in excess of 70 miles per hour.

Tropical storm Isidore occurred during the period between September 25 and October 1, 1984. On the 27th, it had winds of 50 miles per hour. Landfall occurred between Vero Beach and Melbourne on the evening of the 27th. From there it went to Orlando at about midnight, then travelled west, to about 75 miles north of Tampa. On the 28th it made another turn, headed northeast, crossing over to Jacksonville and then out to sea; the storm was accompanied by heavy rains.

Hurricane Bob was relatively short-lived and struck the southwest Florida coast near Ft. Myers on July 21-25, 1985 as a tropical storm. Winds were 50-70 miles per hour. Bob crossed Lake Okeechobee and went out to sea near Vero Beach on the 23rd of July, followed by a sharp turn to the north, skirting Daytona on the 24th. Bob became a hurricane at sea on the 24th, east of Georgia.

On October 9-13th, 1987, Hurricane Floyd appeared. It moved across the western tip of Cuba on the 11th on a northeast track. A more eastern turn was made across the Dry Tortugas and into the Keys on the 12th of October. It became a hurricane near Key West with winds of 80 miles per hour. The eye of the hurricane crossed over Key West at about noon. Warnings were given all across south Florida; some tornadoes occurred in the southwest coast of Florida. The eye of the storm appeared over Marathon later and over Key Largo at about 1800 on the 12th. Floyd's winds were 75 miles per hour and the barometric pressure was 29.32 inches. About 30 miles south of Miami, Floyd broke out into the Atlantic near midnight on the 12th. Winds and rains attributed to Floyd were felt as far as Palm Beach.

Hurricanes Elena, Juan, and Kate, which occurred in 1985, are briefly discussed below.

Hurricane Elena, a Category 3 hurricane in August/September 1985, deserves discussion although it never actually made landfall in Florida (Figure 15). Its center passed within 40 miles of the West Coast, where it stalled for about 24 hours offshore from Cedar Key, and then moved west northwest, passing within 30 miles of Cape San Blas. In its passage the storm tide that was created caused heavy waterfront damage in the City of Cedar Key and the disappearance of 1500 feet of the exposed south tip of Cape San Blas. Because of the offshore location of Elena's peak winds, most of the damage to the coast was due to the storm tide (7-9 feet) and wave activity causing destruction which stretched from Venice to Pensacola. Nearly a million people were evacuated from low lying coastal sections in the warning areas posted for Hurricane Elena (Case, 1986).

While not directly striking the Florida coastline, Juan (Figure 16) in October/November 1985, nevertheless, impacted the extreme northwest Florida panhandle. Also, Pinellas, Manatee, Sarasota and Lee Counties were continuously pounded by the storms spiral bands through the evening of Halloween (see Figures 16a, b and c).

On November 21, 1985, Hurricane Kate (Figure 17), a category 2 storm, struck the coast near Port St. Joe in the Florida Panhandle. It was the only hurricane to strike Florida so late in the season this far north. Just prior to making landfall near Mexico Beach, about halfway between Panama City and Port St. Joe, Kate slowed her forward speed and weakened in the early morning because of cooler sea surface temperatures in the northern Gulf of Mexico. The total damage, was mainly due to the storm tide and wave activity. A sizeable $300 million in damages (adjusted to 1990 dollars) resulted, yet it caused only about one-fourth of the damage inflicted by Hurricane Elena. As with Elena, damage to the coast was mainly due to the storm tide and wave activity.

Hurricane Chris was a tropical storm during the period August 21-29th, 1988. Chris, with heavy rains, skirted the Florida east coast from Miami to Jacksonville, first as a tropical depression then as a tropical storm on the 27th and 28th.

Hurricane Keith was a tropical storm during November 17-24, 1988. The storm moved into Florida's west coast between Ft. Myers and Tampa on the 22nd, a tropical storm with 65 miles per hour. The storm crossed the state intact and came out into the Atlantic near Melbourne and Cape Canaveral on the 23rd. Heavy rains and some tornadoes were sighted throughout the state.

While 1990 produced 14 named storms, the most since naming began in 1953, only Marco, a tropical storm affected the northwest portion of Florida slightly and Klaus, as a final disturbance got into the central and north central part of Florida. 1991 couldn't rally anything more than a brush with tropical storm Fabiàn on the extreme southeast tip of Florida.

Hurricane Bob, the most potential of the 1991 season headed toward the Miami-Palm Beach area but 200 to 300 miles off the Florida coast he executed an almost 90 degree turn to the north and missed all of Florida. So the Sunshine State escaped once again. Thus, in 1991 no hurricanes struck Florida.

Chapter 3

Hurricane Andrew

Except for several tropical depressions, June, July and half of August of the 1992 hurricane season was quiet. The last late start was Anita back in 1977 on the 28th of August, in the Gulf of Mexico.

But on August 14th, 1992, satellite photos indicated a strong tropical wave off the African coast in the area of the Cape Verde Islands. This system moved west for two days and developed into a tropical depression near 11.6N and 40.4W early on the 17th. By noon of the 17th the winds were 40 miles per hour and Tropical Storm Andrew was named. This position was about 1175 miles east of the Lesser Antilles.

By the 20th, Andrew was in trouble, with winds less than 45 miles per hour and the barometric pressure was that of normal sea level; the whole system was shaky. At this point, San Juan, Puerto Rico, was only 350 miles southwest, but Andrew had slowed down!

The next morning, however, winds were up to 60 miles per hour and pressure had dropped to 29.71 inches. By 2300 on the 21st, Andrew was 610 miles east of Nassau, in the Bahamas, with 65 mile per hour winds.

The morning of the 22nd of August, air reconn confirmed that, "Andrew is now a hurricane". Winds were 76 miles per hour, pressure was 29.35 inches and he was 800 miles east of Miami, Florida.

By 2300 on the 22nd Andrew was moving dead west at 15 miles per hour with 110 mile per hour winds and a pressure of 28.32 inches, a Category 2 hurricane.

But by noon of the 23rd we had a Category 4 hurricane! Winds were 135 miles per hour, pressure had dropped to 27.46 inches, and the storm was 330 miles east of Miami, still moving west at 16 miles per hour.

By 1415 that same afternoon, Andrew was at his peak with 150 mile per hour winds and 27.23 inches (Andrew was very close to a Category 5 storm). At this point a Hurricane Watch was posted from Titusville south to Vero Beach and Hurricane Warnings covered from Vero Beach south through the keys and up the west coast to Ft. Myers.

By 2100 on August 23rd, Andrew was in the Bahamas 180 miles

east of Miami. Landfall near Miami was predicted for early morning August 24.

Between 0400-0500 on the 24th, Andrew struck the Florida coastline just south of Miami, with sustained winds of 145 miles per hour and recorded gusts of 164 miles per hour, reported by the National Hurricane Center in Coral Gables, before the main radar at the center was destroyed. Gusts to 175 mph were later confirmed.

Andrew crossed the state with 125 mile per hour winds and a forward speed of 18 miles per hour, still moving dead west, and a Category 3 storm now. Pressure was 27.91 inches. Some recorded gusts in miles per hour were:

Palm Beach International Airport	54
Goodyear Blimp Base at Pompano	100
Miami International Airport	115
National Hurricane Center	163
Turkey Point Power Point	163
Turkey Point Nuclear Power Plant	160
Fowey Rocks (Biscayne Bay)	169

(U.S. Army Corps of Engineers, 1993.)

Once into the warm waters of the Gulf, winds returned to 140 miles per hour, or Category 4 again.

By 0600 on the 25th, Andrew was 270 miles southeast of New Orleans, now moving west-northwest at 17 miles per hour. Winds were 140 miles per hour.

At 1300 on the 25th, the storm was 150 miles south of New Orleans moving west-northwest at 16 miles per hour. Winds were still 140 miles per hour and barometric pressure was 27.85 inches.

The storm slowed-down to almost stationary 30 miles south- east of Lafayette, Louisiana. Early on the 26th of August, winds near new Iberia, Louisiana, were reported to be 115 miles per hour with gusts to 160 miles per hour.

Landfall occurred between New Iberia and Lafayette, Louisiana, as a Category 3 hurricane.

By noon on the 26th Andrew was down-graded to tropical storm status for the first time since the 22nd of August. Near Baton Rouge, Louisiana, there was up to 10 inches of rain and 65 mile per hour winds, with tornadoes.

By the morning of the 28th, the system was in eastern Tennessee, trying to merge with a cold front, the remains of hurricane Lester, a Pacific hurricane. Andrew finally died out in Pennsylvania on August 29, 1993.

On Sabbatical with Hurricane Andrew

After anchoring their 40-foot sailboat (named *Sabbatical*) in Manatee Bay in the upper Keys, Dr. and Mrs. Stephens took refuge in a friend's home in Southwest Miami.

The following is an eye witness account written by Lois Stephens of Melbourne Beach, Florida.

> Sleep was difficult, but I think we all managed to sleep some. About 2:00 AM it started. The wind was howling and shutters were banging. The five of us all crowded into the hallway, just like the usual pre-hurricane instructions stated. Fortunately, Karen had put out candles for us. So far so good. The lights went out, the rain started. The wind got many times stronger and the house almost shivered. The force became so great we ran almost panicky into the bathrooms. There were two, both without windows. Ron and Karen headed for one, Lee, Tom and I the other. We sat on lawn chairs, nestled close together, in the dark with our eyes closed. We opened the door only long enough to get a small votive candle, but the force became too great to open it. The wind grew more ferocious. Suddenly, the windows began to blow out, one at a time, fiercely smashing against the tiled floors. One huge crash I assumed to be the TV, but it was the newly purchased computer. Glass kept smashing. I had been aware for some time of my two root canals. It was strange, but the teeth had piercing pain. I remembered once before being in an airplane with inadequate pressure regulation and experiencing the same pain. Then it hit. The drop in pressure in the house was so intense it caused pain in your ears and you had to keep swallowing, something like when a plane takes off, but much, much worse. We tried to open the bathroom door, but the force was too great. So three and two of us sat in silence, eyes closed, waiting for the horror to end. The small door to the "attic" storage space blew in and the rain followed. Water crept in around our feet, and I had a dread of it rising. But

it did not. Sometime after 6:00, I think, the wind subsided substantially, and we had nerve enough to leave our sanctuary. The house was all but demolished. The bed where Tom and I had slept a few hours before was full of glass and wet soggy debris. (My emergency bag of clothing, etc. was waterproof, but I had left it unzipped so it was likewise wet and full of junk.) The newly tiled (and in 3 rooms, newly carpeted) floors were covered with roof shingles, nails, much glass of all sizes, furniture, books, and of course, with a couple of inches of water. Ceiling fans still clung to their mountings, but under each, the light globes were full of dirty water. Water oozed from holes in the walls where Karen's (she is an artist) newly framed tropical paintings had been hung. Paint was stripped from the walls. The carport (a sturdy "permanent" one) and door overhang were gone. The new roof was without shingles, and had gaping holes. A look outside showed that all trees and fences were down. It was, of course, light now, so being cautious but ignoring some of the warnings we had heard, we walked around the neighborhood. It was sickening, horrifying. Not one house had escaped major damage. Trees, even the largest, were sprawled over houses, cars and streets. Some cars had only broken windows and dents (as did our friend's), and some were blown about and overturned. One had burned from a fallen power line. Not just the power lines were down, but heavy duty power poles were also broken. Except for no smoke or fires at this point, it must have been what a "bombed out" area looks like in wartime. Miraculously, quick checks with neighbors found no one injured. Since roads in every direction were impassable, any hope of getting back to what might or might not be left of our boat were given up for the present.

...LATER...

Highways were somewhat clear by this time, except for some questionable power lines. Trees and large downed poles lined the way. What was most amazing, though, was that literally thousands of cars had found their way to the same area where we were. Traffic was next to impossible, lights and signs inoperative, and cars in extremely questionable condition. We'll never know the number of traffic accidents that day alone.

We passed the hotel, the Holiday Inn, where we had tried so desperately to get a room. It was standing, but barely, with all windows, balconies, etc. blown away. We passed houses with walls only and houses without any walls. Devastation went on for miles. We passed lines of hundreds of people waiting for water. Huge trucks had apparently been placed there at some point to distribute bottled water. One truck had blown uselessly on its side.

Eventually we got to our boat - it was not where we had left it, of course, but it looked good and was tightly nestled back in a grove of mangroves, aground. Miraculously, even the little Zodiac dinghy was still tied to it, snuggled alongside like a loyal puppy nestled against its master. A window was out, glass was everywhere and branches were entwined in some lines. A stanchion (Tom says) was out and leaves and red mud covered one side of the boat. It was beautiful - we were ecstatic. The carpet was wet - the galley was soaked and covered with glass, but everything else was as we left it.

That night, Tom and I were alone in the middle of Manatee Bay, the most beautiful anchorage of our entire sailing experience. The sky was clear and bursting with stars with no electric lights to distract from their beauty. There were no airplanes, distant cars, trains or any noises. The most amazing phenomena was taking place in the water around us. We had seen luminous fish on occasion, but we saw intensely brilliant green fish swimming around the boat. We dropped a line in the water and swirled it around and it left a trail of light behind it, somewhat like a comet. If we splashed the water, we splashed thousands of tiny lights. (All of this, of course, sent us later to our reference books to see what we had discovered.) We were so fortunate, so thankful, and we sipped our champagne.

...THE NEXT DAY...

We were stopped by the Miami Police in a huge inflatable boat and advised we were on the Coast Guard "list" (missing persons and boats) and to "call home".

Chapter 4

Andrew Epilogue

As of this writing, assessment of Hurricane Andrew is incomplete. However, the following is a reasonable preliminary estimate of death and destruction and some important characteristics of the storm.

Current death toll stands as 41. This is far less than what has occurred in past hurricanes of comparable strength.

Hurricane Andrew is the most destructive natural disaster in U.S. history! Damage estimates are fluctuating between $15 and $30 billion, most of which is in southern Dade and Monroe Counties, Florida, from Kendall southward to Key Largo. The Bahamas are estimating at least $250 million dollars in damage and Louisiana more than $1 billion.

Florida's agricultural industry loss was $1.04 billion alone. There was moderate impact damage to the offshore reef areas down to a depth of 75 feet (U.S. Army Corps of Engineers, 1993).

117,000 homes were destroyed or had major damage and 90% of all homes in Dade County had major roof damage (U.S. Army Corps of Engineers, 1993).

According to the U.S. Army Corps of Engineers who worked cooperatively with other agencies to determine environmental impacts, 12.7 million cubic yards of debris resulting from Andrew were hauled away; there were 39 approved debris burning sites (Figure 25).

Damage to the Turkey Point nuclear powerplant belonging to Florida Power and Light Co. was $100 million (U.S. Army Corps of Engineers, 1993).

In terms of damage to moored recreational vessels within Biscayne Bay, a total of 918 hurricane damaged vessels were found. According to Antonini et al. (1993), ..."roughly ... one-third of the damaged vessels were completely or partially submerged, damaged but floating, and damage aground." The site of the greatest devastation was in the area of Dinner Key Marina near Coral Gables in Miami.

Massive evacuations were ordered in Florida and Louisiana. This accounts for the low death rate. It's called Hurricane Preparedness.

The recovery process is still underway (Figures 19 and 20), but it should be emphasized that the results of tremendous structural damage by Andrew's winds could become accumulative in the future.

Andrew was a compact system with a radius of maximum winds of about 12 miles. A slightly larger system or one with a landfall a few miles further north would have been even more catastrophic by affecting the more heavily populated areas of Greater Miami, Miami Beach, and Fort Lauderdale. New Orleans was relatively spared also.

Such statistics as the 16.9 foot storm tide in Biscayne Bay, Miami, is a record maximum for southeast Florida. Louisiana had 7 foot storm tides.

Only Hurricane Camille in 1969 and the "Great Labor Day Hurricane of 1935" in the Florida Keys had lower Barometric pressures at landfall in this century. Barometric pressure associated with Andrew bottomed out at 27.23 inches.

A maximum 10-second flight-level wind speed of 170 knots, or 196 miles per hour, was reported by the reconnaissance aircraft in the vicinity of northern Eleuthera Island in the Bahamas on the 23rd of August. The storm surge there was 23 feet!

Andrew will not be the last hurricane to cause such massive devastation and havoc. Another similar storm may appear next year, or ten years from now—there is no way to know when. However, the bitter lessons we have learned should provide us with ample ammunition to survive the next big one.

The 1993 Hurricane Season

No hurricanes or tropical storms struck Florida or seriously affected Florida in 1993. The most powerful storm of the 1993 season was Hurricane Emily, August 22 to September 6. She was a Category 3 hurricane with top winds of 120 miles per hour and a low pressure of 28.38 inches. This storm came directly at Florida until the 28th of August, at which time she turned to the northwest. Emily, because of Andrew in 1992, did a first-class scare-job on the Florida coast from Miami to Jacksonville but she never got to within 800 miles of the Florida coast at any point. Emily scraped the Cape Hatteras area with minimal damage then turned back east again to die out some 480 miles south, southeast of Cape Race, Newfoundland.

References

Anon., 1926. Tropical Hurricane Spreads Disasters Along East Coast, Melbourne Journal, Melbourne, Florida, July 27, 1926.

Antonini, G.A., P.W. Box, E. Brady, M. Clarke, H.R. Ledesma, and J.L. Rahn. 1993. Location and Assessment of Hurricane Andrew Damaged Vessels on Biscayne Bay and Adjoining Shores. Florida Sea Grant College Program, Gainesville, FL 58 pp.

Anon., 1928. Northern Extremity of Tropical Hurricane Sweeps through Melbourne, Melbourne Times-Journal, Melbourne Florida, August 10, 1928.

Balsillie, J.H. 1985. Post-Storm Report: The Florida East Coast Thanksgiving Holiday Storm of 21-24 November 1984 Department of Natural Resources, Division of Beaches and Shores, State of Florida, Tallahassee, 74 pp.

Bigelow, F.H., 1898. Features of Hurricanes (Originally published in Yearbook of the Department of Agriculture for 1898). Quoted In: West Indian Hurricanes by E.B. Garriott, Weather Bureau, Washington, D.C., 69 pp + tracking charts.

Case, R.A. 1986. Atlantic Hurricane Season of 1985. Monthly Weather Review, Volume 114, No. 7, 1390-1405.

Clark, R.C. 1986a. The Impact of Hurricane Elena and TS Juan on Coastal Construction in Florida. Beaches and Shores Post-Storm Report 85-3. Department of Natural Resources, Division of Beaches and Shores, State of Florida, Tallahassee, 142 pp.

Clark, R.C. 1986b. Hurricane Kate - Beaches and Shores Post-Storm Report 86-1. Department of Natural Resources, Division of Beaches and Shores, State of Florida, Tallahassee, 114 pp.

Dunn, G.E. and B.I. Miller. 1964. Atlantic Hurricanes. Louisiana State University Press, Baton Rouge, 377 pp.

Dunn, G.E. and Staff 1967. Florida Hurricanes. Technical Memorandum WBTM SR-38, Environmental Sciences Services Administration, National Hurricane Center, Coral Gables, Florida.

Frank, N., 1978. Hurricanes in Brevard County, Florida Today, Melbourne, Florida, Sunday, June 4, 1978.

Garriott, E.B., 1900. West Indian Hurricanes. Weather Bureau, Washington, D.C.

Gray, R. W. 1949. Florida Hurricanes, In: Monthly Weather Review, Volume 61, No. 1, January 1933. (Revised by G. Norton and reprinted as a separate pamphlet, 1949, 6 pp.)

Gray, W. M., 1990. Strong Association Between West African Rainfall and U.S. Landfall of Intense Hurricanes. Science, Volume 249, pp 1251-1256.

Hebert, P.J., J.D. Jarrell, and M. Mayfield. 1992. The Deadliest, Costliest, and Most Intense United States Hurricanes of this Century (and other frequently requested hurricane facts). NOAA Technical Memorandum NWS HNC-31, National Oceanic and Atmospheric Administration, National Weather Service, National Hurricane Center, Coral Gables, Florida, 39 pp.

Holmes, G. W. 1876. Letter to a friend: F. A. Hopwood, Personal collection, Melbourne, Florida, 1985

National Climatic Center. 1954-1979. North Atlantic Tropical Cyclones Series (1954-1979). Climatological Data, National Summary Volumes, National Climatic Center, Asheville, North Carolina, each year paginated.

National Oceanic and Atmospheric Administration (NOAA), 1970-1979. Climatological Data, National Summary, Volume 21-30. No. 13 National Climatic Center, Asheville, North Carolina, unpaginated.
NOAA, 1982. Some Devastating North Atlantic Hurricanes of the 20th Century. U. S. Department of Commerce, NOAA, Washington, D.C., 14 pp.

NOAA, 1987. Tropical Cyclones of the North Atlantic Ocean, 1871-1986. Historical Climatology Series 6-2, National Climatic Center, Asheville, North Carolina, 186 pp.

NOAA, 1993. "Hurricanes" A Familiarization Booklet, Revised April 1983, NOAA PA 91001, National Oceanic and Atmospheric Administration, National Hurricane Center, Coral Gables, Florida, 36 pp.

Rabac, G. 1986. The City of Cocoa Beach, the First Sixty Years. Apollo Books, Winona, Minnesota, Page vii.

Simpson, R.H. and H. Riehl. 1981. The Hurricane and Its Impact. Louisiana State University Press, Baton Rouge and London, 399 pp.

Sugg, A.L. 1966. The Hurricane Season of 1965. Monthly Weather Review, Volume 94, No.3, pp 183-191.

Sugg, A.L., L.G. Pardue, and R.L. Carrodus. 1971. Memorable Hurricanes of the United States since 1873. NOAA Technical Memorandum NWS SR-56, National Oceanic and Atmospheric Administration, National Weather Service, Southern Region, Fort Worth, Texas, 52 pp.

Tannehill, I. R. 1938. Hurricanes, their Nature and History. Princeton University Press, Princeton, New Jersey, 304 pp.

U.S. Weather Bureau, Monthly and Annual Reports, 1897-1965. Clima-to logical Data, Florida, April 1897-December 1965. National Climatic Center, Asheville, North Carolina, microfiche 112 fiche.

U.S. Army Corps of Engineers. 1993. Hurricane Andrew Storm Summary and Impacts on the Beachs of Florida, Special Report. Jacksonville District, Florida, 61 pp. plus several appendices.

Yanaros, J., 1986. Personal Communication. Damage Caused by Hurricane in Ft. Pierce, 1933, Melbourne, Florida.

Tables, Figures and Plates

Table 1. Saffir-Simpson Scale for Classifying Hurricanes

Cate-gory	Pressure (millibars)	(inches)	Winds (mph)	Surge (feet)	Damage
1	980	28.94	74-95	4-5	minimal
2	965-975	28.50-28.91	96-110	6-8	moderate
3	945-964	27.91-29.47	111-130	9-12	extensive
4	920-944	27.17-27.88	131-155	13-18	extreme
5	≤920	≤27.17	≥155	18	catastrophic

Table 2. Number of hurricanes, tropical storms and combined total storms by 10-year periods.

10-Yr Average	Hurricanes	Tropical Storms	Total
1871-1880	NA	NA	21
1881-1890	NA	NA	21
1891-1900	10	11	21
1901-1910	6	11	17
1911-1920	7	4	11
1921-1930	9	2	11
1931-1940	6	12	18
1941-1950	12	8	20
1951-1960	3	10	13
1961-1970	7	4	11
1971-1980	3	1	4
1981-1992	4	8	12
TOTAL	74	76	180*

*Includes 30 tropical cyclones of unknown intensity.

Table 3. The table below is presented here for reference only. It was used to classify intensities of hurricanes through about 1970 and was replaced by the Saffir-Simpson Scale. (Adapted from Dunn and Miller, 1964)

Hurricane Classification in Use Prior to 1970

Hurricane Intensity	Maximum Winds (mph)	Minimum Central Pressure (inches)
Minor	74	29.40
Minimal	74-100	29.03-29.40
Major[1]	101-135	28.01-29.00
Extreme[2]	136	28.00

[1] Major Hurricane: A hurricane with winds 111 mph or more also referred to as a category 3 or higher hurricane as classified by the Saffir-Simpson Scale. Prior to 1970, this description was used.

[2] Extreme Hurricane: An extreme hurricane, also called a "Great" hurricane, is a hurricane of great intensity (winds 125 mph or more) and great size (diameter of hurricane wind 100 miles or more) and other factors such as minimum pressure, storm tides, destruction, and fatalities (Gray as revised by Norton, 1949). The term "Great Hurricane" was used to classify storms through 1970) and is not used any more today. Norton (1949) classified 10 hurricanse as "Great" between 1880 and 1948 and are listed below:

August 1880:	Palm Beach - Lake Okeechobee
June 1886:	Apalachicola - Tallahassee
October 1890:	Key West - Fort Myers
July 1916:	Pensacola - Mobile
September 1919:	Key West
September 1926:	Miami - Pensacola
September 1928:	Palm Beach - Lake Okeechobee
September 1935:	Great Labor Day Hurricane, Florida Keys
October 1944:	Key West - Tampa - Jacksonville
September 1947:	Fort Lauderdale - Fort Myers

Table 4. Chronological list of hurricanes in Florida. (Only hurricanes are listed.)

Date	Name	Area affected	Peak winds MPH	Min. pres. inches	Max. surge feet	Damage - Death - Other Data
1871 Aug.	Unk	Cocoa Bch	Unk	Unk	Unk	Unk, Reference: Frank (1978). Not confirmed as a full hurricane. Direct hit, East-Central FL.
1873 Oct.	Major	20 miles SSE of Venice	Unk	Unk	14	Unk, Reference: Dunn and Miller (1960). Punta Rassa, FL destroyed. From west coast, across state to direct hit East-Central FL.
1876 Sept.	Unk	Eau Gallie (Melb.)	Unk	Unk	Unk	Unk, Reference: Frank(1978) and Holmes (1876). Indirect hit East-Central FL.
1880 Aug.	Major	Cocoa Bch.	Unk	Unk	Unk	Unk, Reference: Frank (1978) and Gray (1949). Severe damage at Palm Bch and Lake Okeechobee. Direct hit East-Central FL.
1885 Aug.	Unk	20 miles E.of Cocoa Bch	Unk	Unk	Unk	Unk, Reference: Rabac (1986). Indirect hit East--Central FL.
1886 June	Major	Apalachicola	Unk	Unk	Unk	Unk, Reference: Gray (1949), NOAA (1987). High tides.

Date	Name	Area affected	Peak winds MPH	Min. pres. inches	Max. surge feet	Damage - Death - Other Data
1886 July	Unk	St. Marks	Unk	Unk	Unk	Unk, Reference: NOAA (1987).
1886 July	Unk	N. of Cedar Key	Unk	Unk	Unk	Unk, Reference: NOAA (1987).
1887 July	Unk	Valparaiso	Unk	Unk	Unk	Unk, Reference: NOAA (1987).
1888 Aug.	Unk	Miami	Unk	Unk	14	Unk, Reference: NOAA (1987).
1888 Oct.	Unk	Cedar Key	Unk	Unk	Unk	9 died, Reference: Dunn and Miller (1960), NOAA (1987).
1889 Sept.	Unk	Pensacola	Unk	Unk	Unk	Unk, Reference: NOAA (1987).
1891 Aug.	Unk	Miami	Unk	Unk	Unk	Unk, Reference: NOAA (1987).
1893 June	Unk	Cross City	Unk	Unk	Unk	Unk, Reference: NOAA (1987).

1894 Sept.	Major	Key West, Punta Gorda	104 Key West	Unk	Unk	Unk, Reference: Dunn and Miller (1960), NOAA (1987).
1894 Oct.	Unk	Apalach-icola	Unk	Unk	Unk	Unk, Reference: NOAA (1987).
1896 July	Major	Pensacola	100	Unk	Unk	Unk, Reference: Dunn and Miller (1960), NOAA (1987).
1896 Sept.	Major	Cedar Key	Unk	Unk	10 Cedar Key	100 died FL. Reference: Dunn and Miller (1960), NOAA (1987).
1896 Oct.	Unk	Punta Gorda	Unk	Unk	Unk	Unk, Reference: NOAA (1987). 68 dead-over 2 million dollars damage. From west coast across state to direct hit East-Central FL.
1898 Aug.	Unk	Apalach-icola	Unk	Unk	10.8 Fern-andina Bch	100 thousand dollars, 12 died. Reference: USWB (1898).
1898 Oct.	Unk	Fernan-dina	Unk	28.95	Mar 4	500 thousand dollars. Reference: USWB (1898).
1899 Aug.	Cat. 2	Carabelle	Unk	28.90	3-4	500 thousand dollars, 6 died. Reference: USWB (1899). 7 ships wrecked.

Date	Name	Area affected	Peak winds MPH	Min. pres. inches	Max. surge feet	Damage - Death - Other Data
1903 Sept.	Cat. 1	Jupiter, Apalachicola	78	29.46	8-10	500 thousand dollars, 14 died, ship wrecked near Jupiter. Reference: USWB (1903).
1906 Sept.	Cat. 2	Pensacola	100	28.29	10	3-4 million dollars, 34 died, Pensacola; 134 total died. Reference: USWB (1906).
1906 Oct.	Cat. 2	Key West, Miami	Unk	28.55 Miami	Unk	160 thousand dollars, 164 died (railroad workers), Miami. Reference: USWB (1906)
1909 Oct.	Cat. 3	Key West	Unk	28.36 Marathon	Unk	1 million dollars, 15 died. Reference: USWB (1906).
1910 Oct.	Cat. 3	Key West, Fort Myers	125 Sand Key	28.20 Fort Myers	15 Key West	365 thousand dollars, 30 died. Reference: Tannehill (1938), Dunn and Miller (1960). This hurricane did a loop in the Gulf of Mexico.
1911 Aug.	Cat. 1	Pensacola	Unk	Unk	Unk	Unk, Reference: USWB (1911).
1915 Sept.	Cat. 1	Apalachicola	Unk	29.25	Unk	100 thousand dollar, 21 died, wrecked sponge vessels. Reference: USWB (1915).

1916 July	Cat. 2	North-west FL	104	29.31	5 Ft.	1 million dollars, 4 died, crop damage. Reference: USWB (1916).
1916 Oct.	Cat. 2	North-west FL	120	28.76	Unk	100 thousand dollars, tower blown down at Pensacola. Reference: USWB (1916).
1917 Sept.	Cat. 3	Pensacola	125	28.29	7.5	Unk. Reference: USWB (1917) and NWS NHC-31 (1992).
1919 Sept.	Cat. 4	Key West	115 Key West	27.37 Dry Tortugas	Unk	2 million dollars, more than 600 died, Key West-anemometer destroyed. Reference: USWB (1919), NWS-NHC-31 (1992). 300 of the deaths were in Key West.
1921 Oct.	Cat. 3	Tarpon Springs	100 Tarpon Spr.	28.11 Tarpon Springs	10.5 Tampa	3 million dollars, 6 died, highest surge since 1848. Reference: USWB (1921). From west coast across state to Ponce de Leon Inlet. Indirect hit East-Central FL.
1924 Sept.	Cat. 1	Port St. Joe	Unk	29.12	Unk	275 million dollars. Reference: USWB (1924).
1924 Oct.	Cat. 1	Marco Island	90	28.80	Unk	Unk. Reference: USWB (1924).
1925 Nov.	Cat. 1	Sarasota, Tampa	Unk	29.50	Unk	1.6 million dollars, 50 died. Latest storm to strike U.S. Reference: USWB (1925). Only one hurricane and one tropical storm in 1925.

Date	Name	Area affected	Peak winds MPH	Min. Pres. inches	Max. surge feet	Damage - Death - Other Data
1926 July	Cat. 2	Jupiter, Indian, River Lagoon	90	28.80	Unk	3 million dollars. Reference: USWB (1926) and Melbourne Times (Anon., 1926). Direct hit East--Central FL.
1926 Sept.	Cat. 4	Miami, Pensacola	138 Miami	27.61 Miami	13.2 Miami	1.4 billion dollars (1990), 243 died, FL. Reference: NHC-31 (1992).
1928 Aug.	Cat. 2	Stuart, Indian, River Lagoon	Unk	28.84	Unk	250 thousand dollars, 2 died. Reference: USWB (1928) and Melbourne Times (Anon., 1928). Direct hit East-Central FL.
1928 Sept.	Lake Okee-chobee Cat. 4	Palm Bch, Lake Okee-chobee	100+	27.43	10-15	26 million dollars (1990), 1836 died. Reference: USWB (1928), NWS NHC-31.
1929 Sept.	Cat. 3	Mara-thon, Panama City	150	27.99	9	821 thousand dollars (1990), 3 died, FL. Reference: USWB (1929), NWS NHC-31.

1933 Sept.	Cat. 3	Jupiter	125	27.98	Unk	4 million dollars (1990), 2 died. Reference: USWB (1933), NWS NHC-31.
1935 Sept.	Great Labor Day Hurricane Cat. 5*	Long Key	200-250	26.35 Record, in this hemis-phere.	20+	6 million dollars, 408 died. *First Cat. 5 on record to strike U.S., first and only CAT. 5 to strike FL. Reference: USWB (1935), NWS NHC-31 and Gray (1949).
1935 Nov.	Yankee Hurricane Cat. 2	Miami	75 Miami	28.73 Miami	6	5.5 million dollars, 19 died. Reference: USWB (1935) and Gray (1949).
1936 July	Cat. 3	Fort Walton Bch	125	28.46	6	200 thousand dollars, 4 died. Reference: USWB (1936), NWS NHC-31.
1941 Oct.	Cat. 2	Miami, Carabelle	123 Miami	28.48	8	700 thousand dollars, 5 died. Reference: USWB (1941). This hurricane looped in the Atlantic.
1944 Oct.	Cat. 3	Sarasota	163	28.02 Dry Tortu-gas	12.3 Jackson-ville	570,150,000 dollars (1990), 18 died in U.S. Refer-ence: USWB (1944) and Gray (1949).
1945 Sept.	Cat. 3	Home-stead	196 Home-stead	28.08	13.7	500,000,000 dollars (1990), 4 died FL. Reference: USWB (1945).
1946 Oct.	Cat. 1	Braden-ton	80	28.95	6	7 million dollars. Reference: USWB (1946).

Date	Name	Area Affected	Peak winds MPH	Min. pres. inches	Max. surge feet	Damage - Death - Other Data
1947 Sept.	Cat. 4	Pompano Bch	155* Hillsboro	27.76	21.6 Clewiston	704,000,000 dollars (1990), 51 died. Reference: USWB (1947) and Gray (1949). * Record, Recorded Wind Speed, FL to date.
1947 Oct.	Cat. 1	Cape Sable	95	28.76	Unk	20 million dollars, 1 died. Reference: USWB (1947).
1948 Sept.	Cat. 3	Key West, Everglades City	122	28.45	19 Ft. Canal Point	18 million dollars, 3 died. Reference: USWB (1948).
1948 Oct.	Cat. 3	FL Keys & Homestead	100	28.44	6.2 Homestead	5.5 million dollars. Reference: USWB (1948).
1949 Aug.	Cat. 3	West Palm Bch, Stuart, Lake Okeechobee	153	28.17	24 Belle Glade	52 million dollars, 2 died. Reference: USWB (1949).

1950 Sept.	Easy Cat. 3	Cedar Key	125	28.29	6.50	3.3 million dollars, 2 died, 38.7 inches rain in Yankeetown- Double loop in Gulf of Mexico just off Cedar Key.
1950 Oct.	King Cat. 3	Miami	150	28.20	19.3 Clewis-ton	28 million dollars, 3 died. Reference: USWB (1950). Indirect hit East-Central FL.
1953 Sept.	Florence Cat.1	Fort Walton Bch	87	Unk	Unk	200 thousand dollars. Reference: USWB (1953).
1956 Sept.	Flossy Cat. 1	Fort Walton Bch	98	28.93	6.10	25 million dollars, 15 died. Reference: USWB (1956) , Dunn and Miller (1960). Possibly formed from storm in Pacific. 3 tornadoes in FL.
1960 Sept.	Donna Cat. 4	Som-brero Key, Fort Meyers	180-200 FL Keys	27.46	13 Fla-mingo 13.7 Tavenier	1,784,070,000 dollars (1990), 50 died, FL. Refer-ence: USWB (1960) and NWS NHC-31. Indirect hit East-Central FL.
1964 Aug.	Cleo Cat. 2	Miami, Ft. Lauder-dale, E. Coast U.S.	138	28.5	6	Almost 600 million dollars (1990), 3 died., 214 Caribbean. This was a Cape Verde, 'text-book' storm. Reference USWB (1964). Eye, 8 to 16 miles in diameter near Miami. Tornadoes re-ported from Davie to Daytona Bch. Indirect hit East-Central FL.

Date	Name	Area affected	Peak winds MPH	Min. pres. inches	Max surge feet	Damage - Death - Other Data
1964 Sept.	Dora Cat. 2	St. Augustine, North FL	125	28.52	12	Over one billion dollars (1990), 5 died. Reference USWB (1964). 10.7 inches rain in Gainesville. Tides over 10 feet at Fernandian Bch. Rains continued in some areas for 4 days. 18.6 inches of rain at Live Oak; 23.7 inches at Mayo.
1964 Oct.	Isbell Cat. 1	SW, SE, Central FL	90	28.47	Unk	Small storm spawning many tornadoes in FL , 1 died. Reference USWB (1964). At least 11 tornadoes in SE FL coast. Vegetable crop heavily damaged.
1965 Sept.	Betsy Cat.3	South FL, Keys, La.	165	27.82	9	Over 6.4 billion dollars (1990), 75 died. Very erratic course through Atlantic Ocean and FL Straits. Reference: USWB (1965). The eye of this hurricane was 40 miles in diameter at one time. 11.8 inches of rain at Plantation Key.
1966 June	Alma Cat. 2	FL Panhandle, SE U.S.	125, Dry Tortugas	28.65	10	Nearly 10 million dollars (1990) , 8 died, FL. Earliest storm, on record to hit U.S. Reference: USWB (1966). 7.7 inches rain in Miami, 6 Ft. tides at St Marks.

1966 Oct.	Inez Cat. 3	South FL, Keys, Mexico	165, Big Pine Key	27.38	15.5	Over 5 million dollars (1990), 48 died. Another storm with an erratic course. Reference: USWB (1966)
1968 Oct.	Gladys Cat.2	NE FL, Cedar Key	90	28.49	6.5	7 million dollars, FL. Reference: ESSA (1968). 6.6 inches rain at Daytona Bch. 5 Ft. tides at Tampa.
1972 June	Agnes Cat. 1	Port St. Joe	86	28.85	7	6 billion dollars (1990), 122 died, FL. Reference: (NWS NHC-31. More than 1000 mile diameter circulation. Spawned 15 tornadoes in FL. 8.5 inches rain at Key West. 12.7 inches of rain at Big Pine Key.
1975 Sept.	Eloise Cat. 3	Mid-way between Ft Walton Bch & Panama City	155	28.20	18	1,081,854,000 dollars (1990), 9 died. Reference: NWS NHC-31.

Date	Name	Area affected	Peak winds MPH	Min. pres. inches	Max. surge feet	Damage - Death - Other Data
1979 Sept.	David Cat. 2, Cat. 4 in Caribbean	Jupiter, Indian River Lagoon, Vero Bch, Melbourne, Fort Lauderdale, Pompano	172	27.28	3-5	487,366,000 dollars (1990), 5 died in U.S., 7 in Puerto Rico, 1,200 in Dominican Republic. Reference: NWS NHC-31. Direct hit East-Central FL as Cat. 2.
1985 Aug. Sept.	Elena Cat. 3	No landfall, closest point Cedar Key and Cape San Blas	125 (aircraft)	28.17	8	1,392,693,000 dollars (1990), deaths unknown. Reference: NWS NHC-31. 11.3 inches rain Apalachicola. 1 million people were evacuated from affected areas. No landfall in FL.
1985 Oct.	Juan Cat. 1	FL, Louisiana, Alabama	85	28.67	Unk	1,635,000,000 dollars (1990) NWS NHC-31.

1985 Nov.	Kate Cat. 2	Mexico Bch, FL and Panama City	135	28.14	8	300 million dollars (1990), 5 died. First SE FL warnings since INEZ (1966) on 18th and 19th November. Reference, In the Monthly Weather Review in 1985 and in NHC.
1987 Oct.	Floyd Cat. 1	Key West, Key Largo, Marathon	80	29.32	Unk	Unk. Reference, Monthly Weather Review (1987) and NHC.
1992 Aug.	Andrew Cat. 4	Bahamas, S. FL, Louisiana	175	27.23	16.9 Miami	15-30 billion dollars (1993) , 48 died, most destructive natural disaster in U.S. history. Reference: NOAA, 1993. Third lowest barometric pressure at landfall in U.S.

Figure 1. Home in Coconut Grove, Miami, September, 1926 Hurricane (Courtesy National Hurricane Center).

Figure 2. Meyer-Kiser Building, NE First Street, Miami, 1926 Hurricane. Building had to be torn down (Courtesy National Hurricane Center).

Figure 3. Sunken boat in Miami, September, 1926 Hurricane. Boat was once owned by the Kaiser Wilhelm of Germany (Courtesy National Hurricane Center).

Figure 4. Damage in Palm Beach, 1928 Hurricane (Courtesy National Hurricane Center).

Figure 5. Destruction in West Palm Beach, 1928 Hurricane (Courtesy National Hurricane Center).

Figure 6. Train blown off track in 1935 Great Labor Day Hurricane, Florida Keys (From News/Sun-Sentinel).

Figure 7. a. (top) Monument to 1935 Hurricane, Islamorada, Florida Keys.
b. Inscription plaque commemorating those who died in the 1935 Hurricane.

Figure 8. Hurricane Donna. Even though Hurricane Donna did not strike Miami, this photograph shows typical damage along Dade County shoreline (Courtesy of National Hurricane Center).

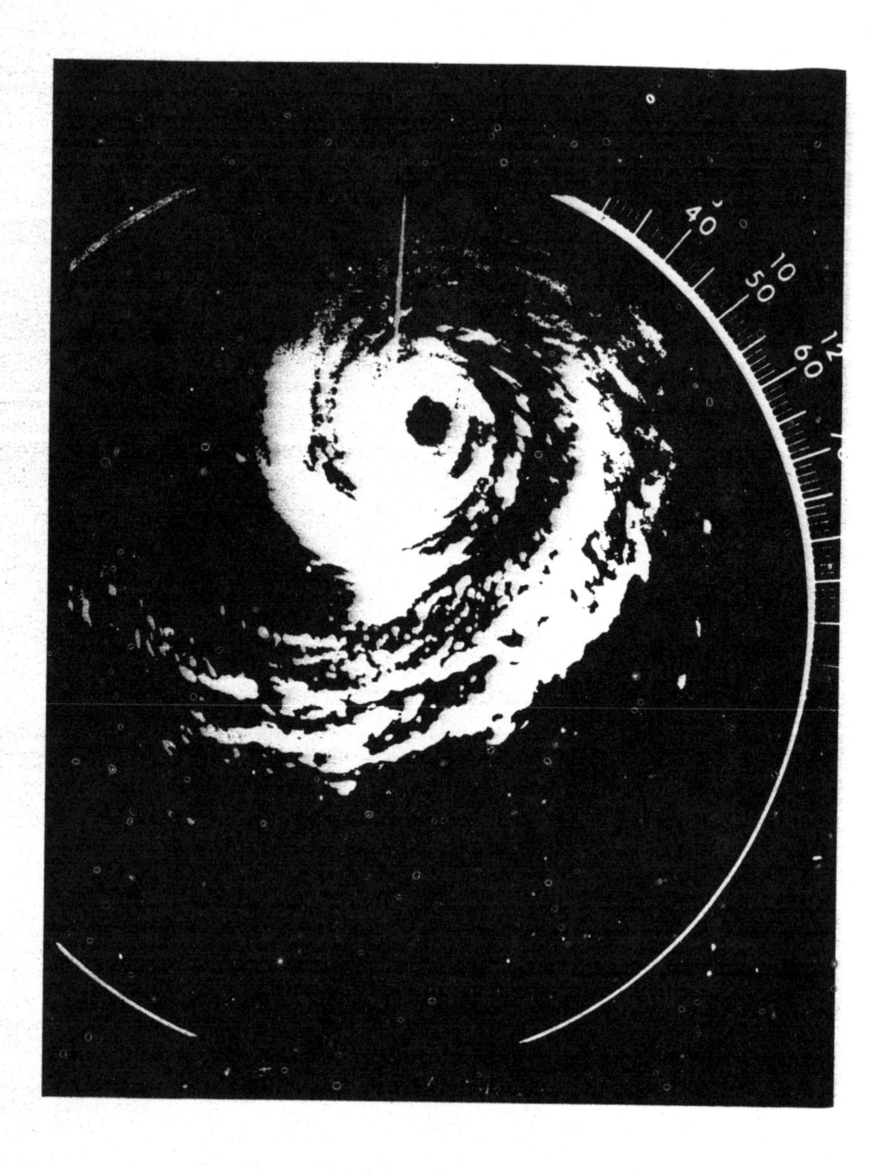

Figure 9. Radar of Hurricane Donna (Courtesy of National Hurricane Center).

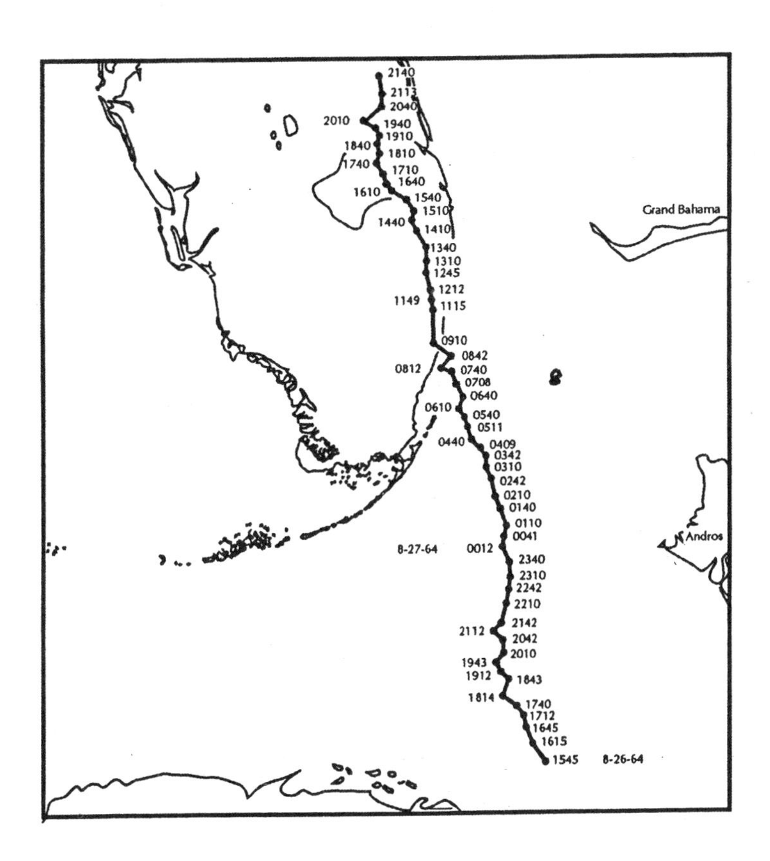

Figure 10. Track and time of Hurricane Cleo in 1964 (from Dunn and Staff, 1967).

Figure 11. 27th Street, Miami, Hurricane Betsy in 1965 (Courtesy of Miami Herald).

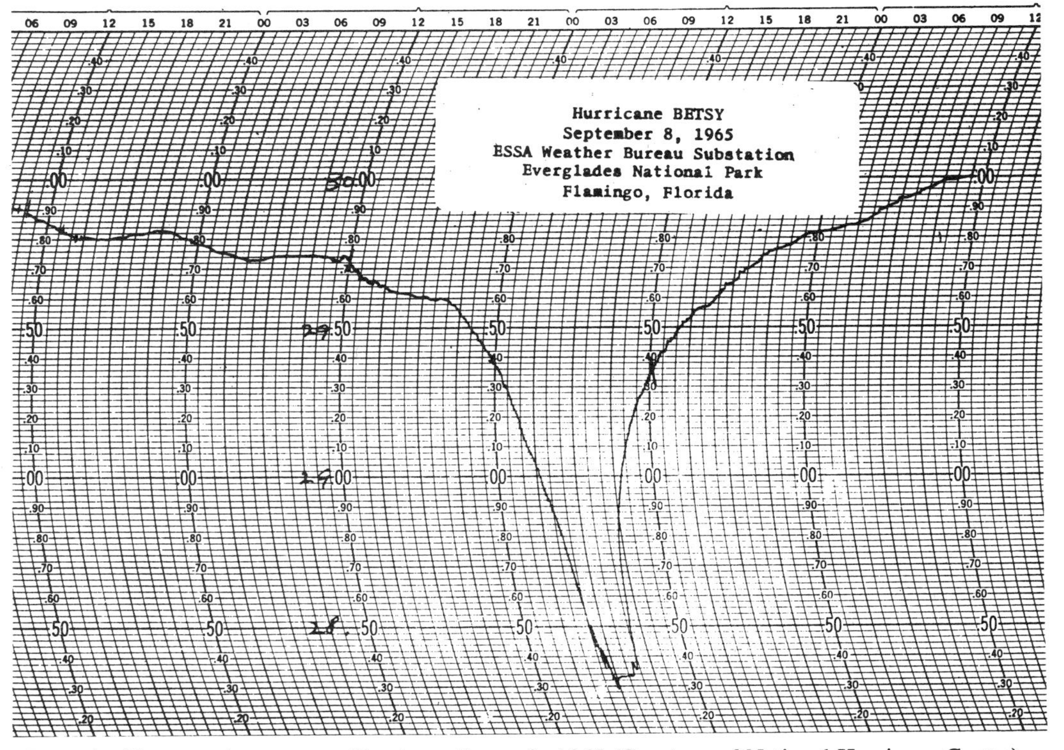

Figure 12. Record of barometric pressure, Hurricane Betsy in 1965 (Courtesy of National Hurricane Center).

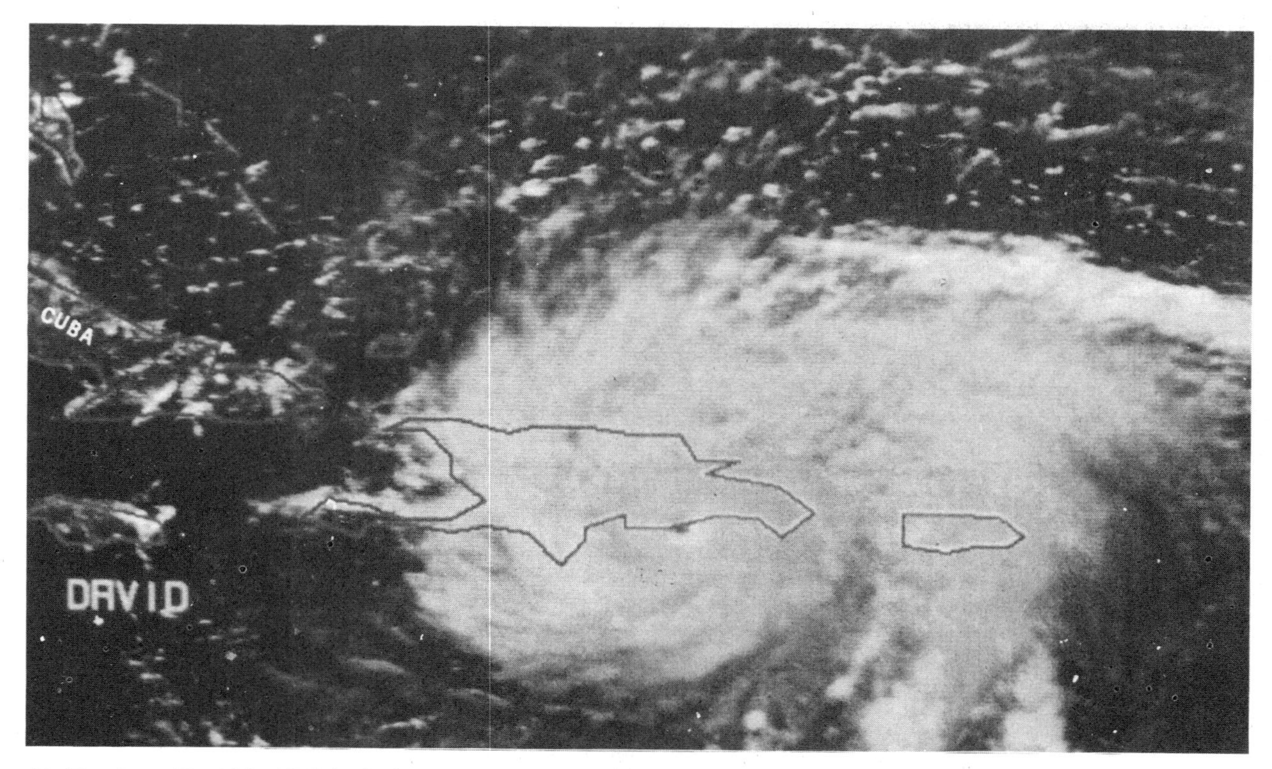

Figure 13. Hurricane David in 1979 in Caribbean (Courtesy of Henry Brandli).

Figure 14a. Hurricane David in 1979 in Florida (Courtesy of Henry Brandli).

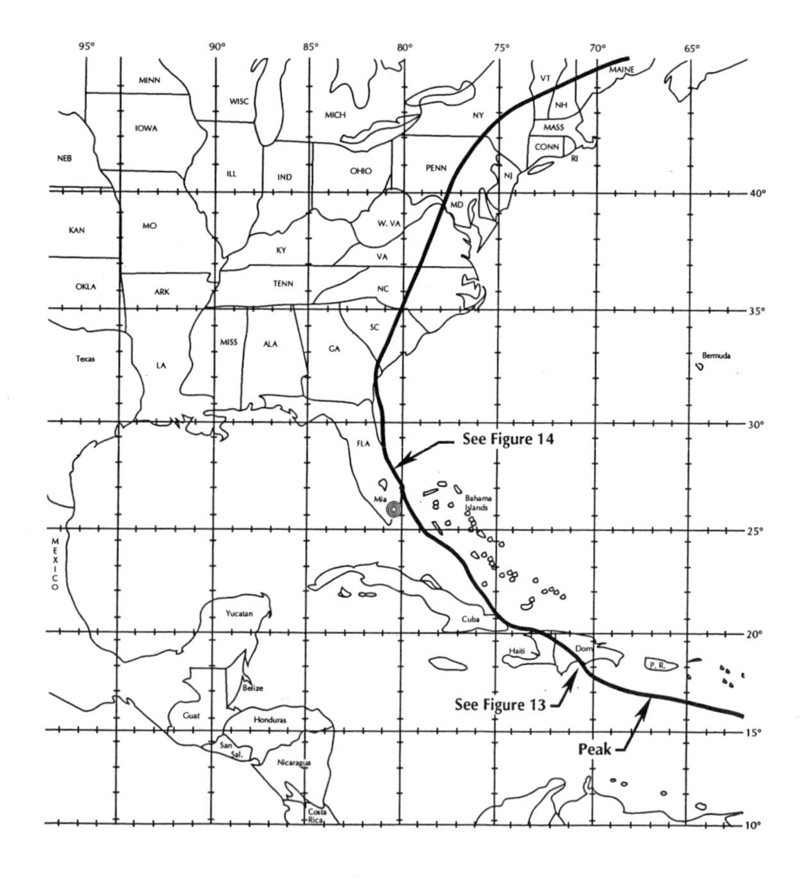

Figure 14b. Hurricane David track (From collection belonging to John M. Williams).

Figure 15. Hurricane Elena (1985) damage (From Clark, 1986a).

Figure 16a. Hurricane Juan (1985) damage (From Clark, 1986a).

Figure 16b. Hurricane Juan (1985) damage (From Clark, 1986a).

Figure 16c. Hurricane Juan (1985) damage (From Clark, 1986a).

Figure 17a (top) and 17b. Hurricane Kate (1985) damage (From Clark, 1986b).

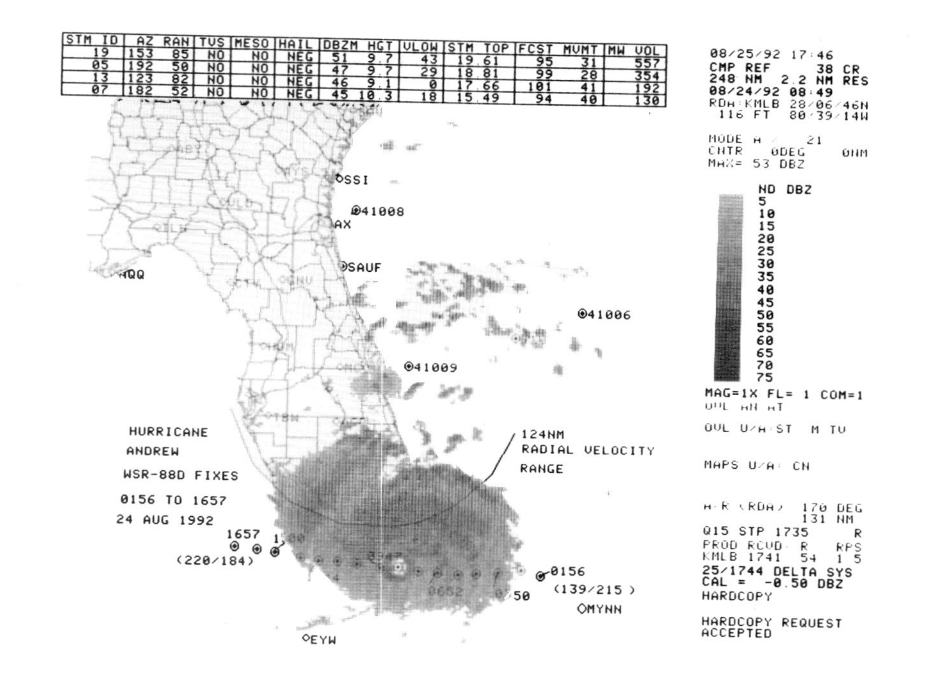

Figure 18. Doppler Radar Image of Hurricane Andrew, 1992 (Courtesy of National Weather Service, Melbourne, Florida Office).

Figure 19. Business sign along US 1 in Homestead, Florida.

Figure 20a (top) and 20b. 20a. Last Chance Saloon, which survived several hurricanes such as Donna, Cleo, Betsy, Inez and Andrew, located on US 1 near Homestead, Florida. 20b. Trees blown down just south of the Last Chance Saloon. Before Andrew, the area around the saloon was heavily wooded.

Figure 21. Typical debris scene from Andrew (U.S. Army Corps of Engineers, 1993).

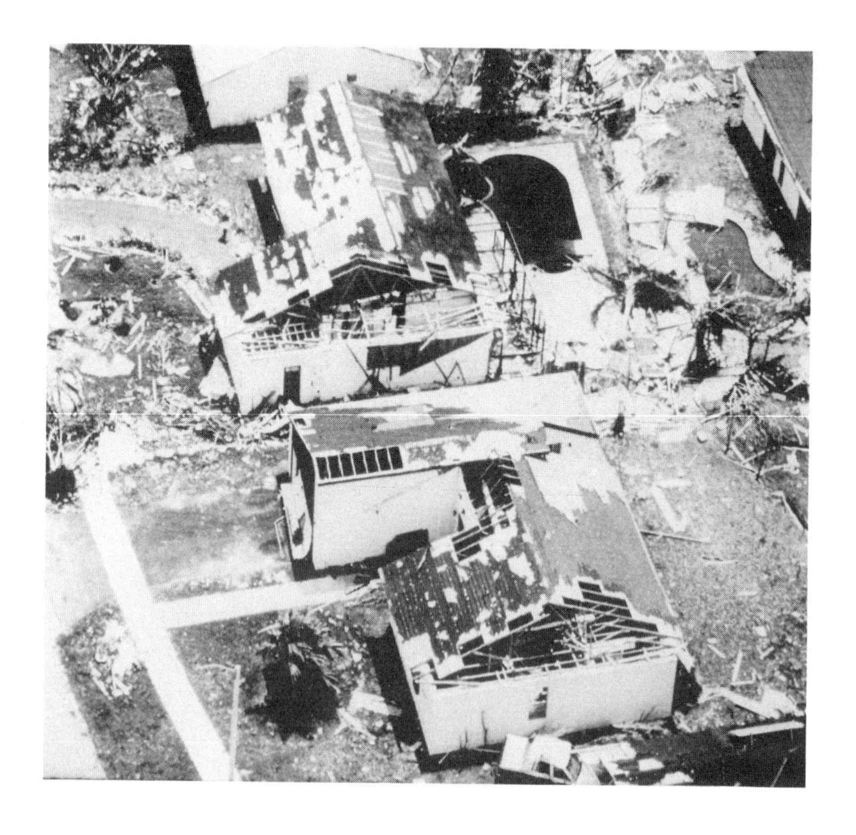

Figure 22. Family home damage from Andrew (U.S. Army Corps of Engineers, 1993).

Figure 23. The mobile home and the hurricane (U.S. Army Corps of Engineers, 1993).

Figure 24. The roof and the hurricane (U.S. Army Corps of Engineers, 1993).

Figure 25. One of 39 debris sites for Hurricane Andrew (U.S. Army Corps of Engineers).

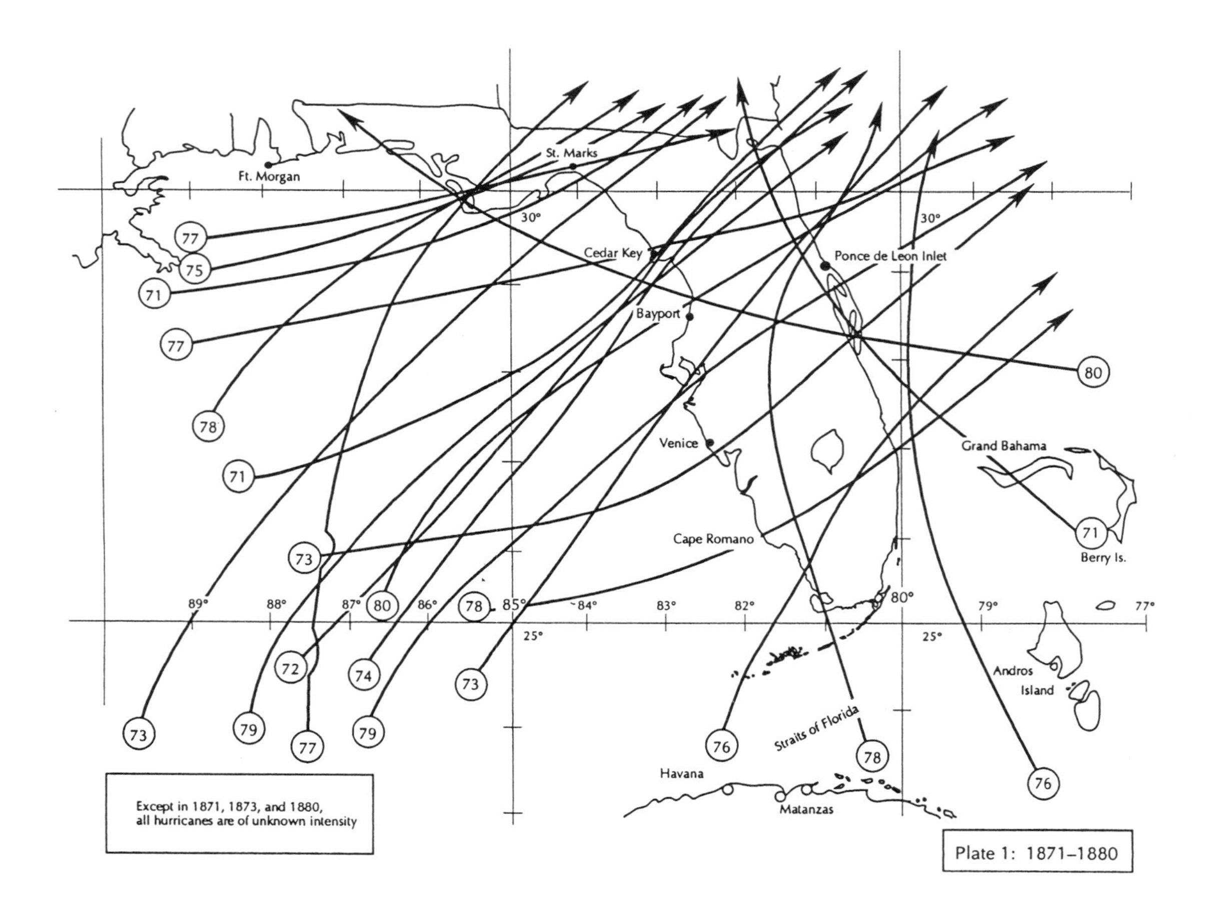

Plate 1: 1871–1880

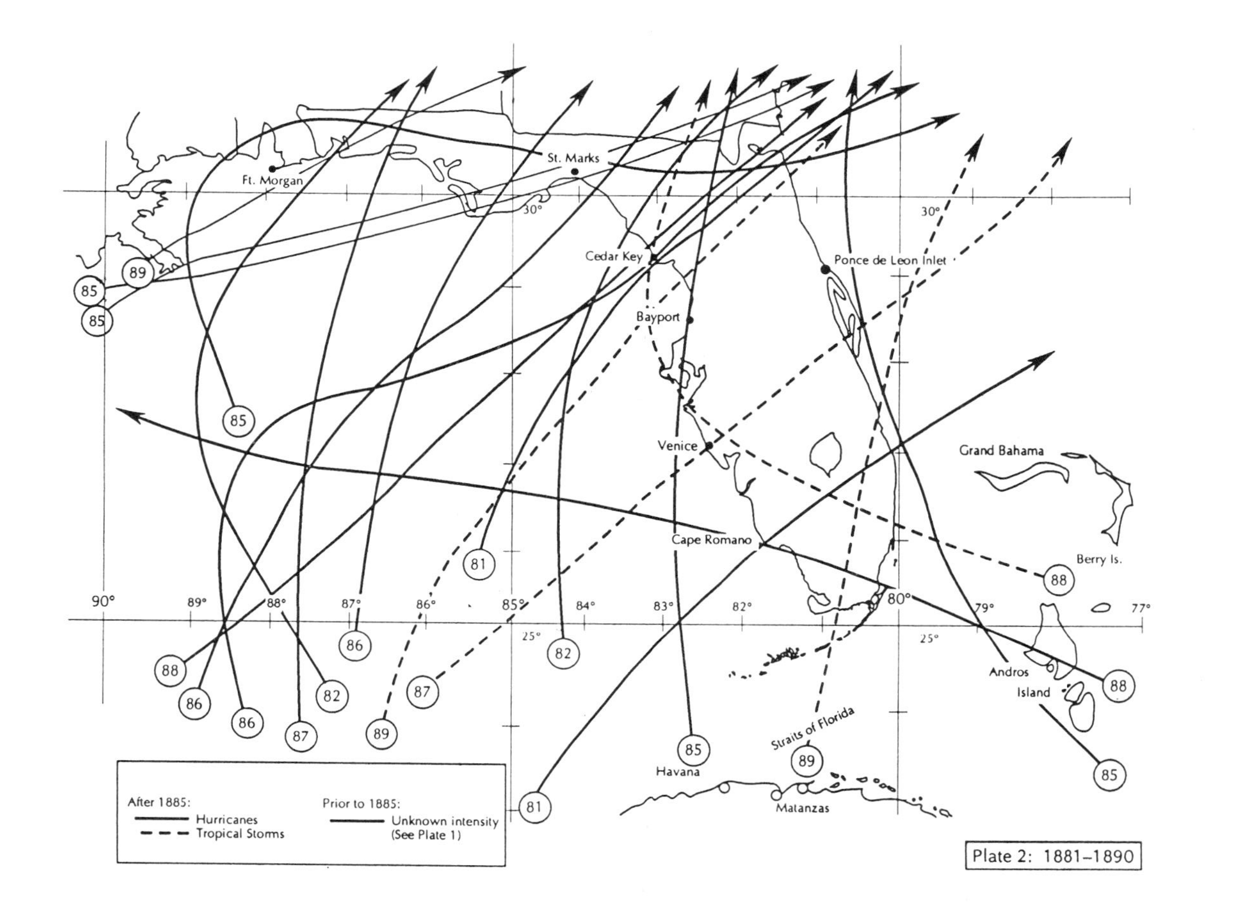

Plate 2: 1881–1890

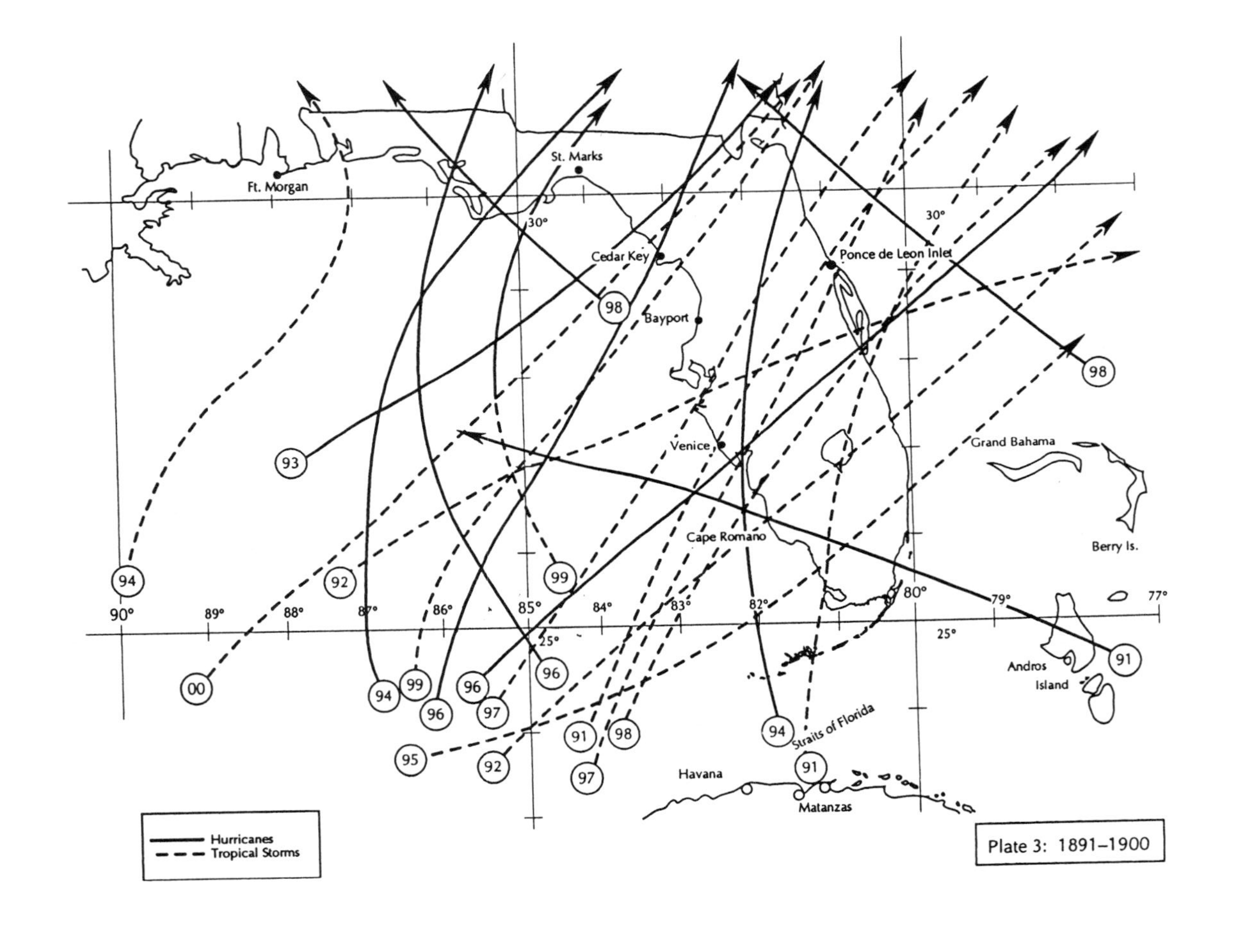

Plate 3: 1891–1900

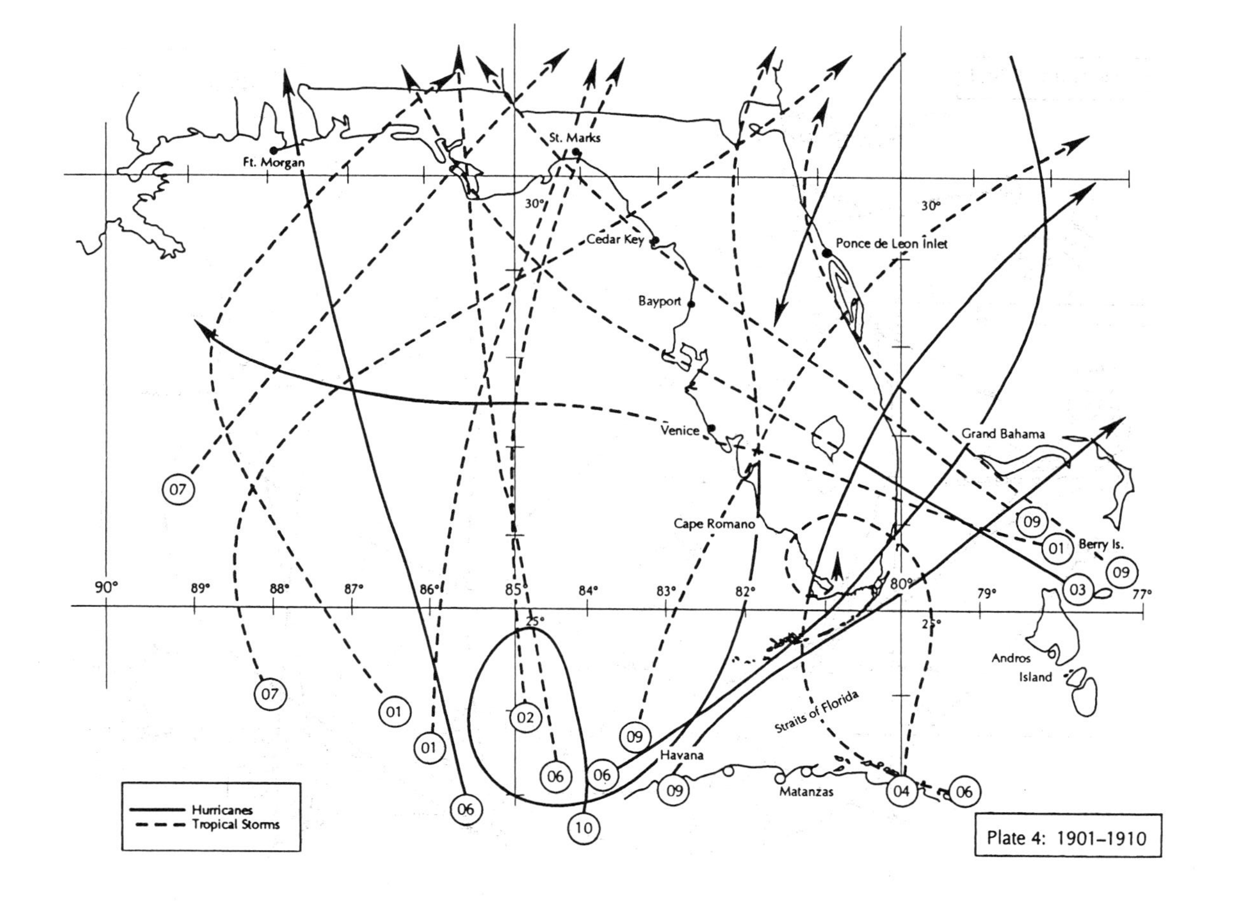

Plate 4: 1901–1910

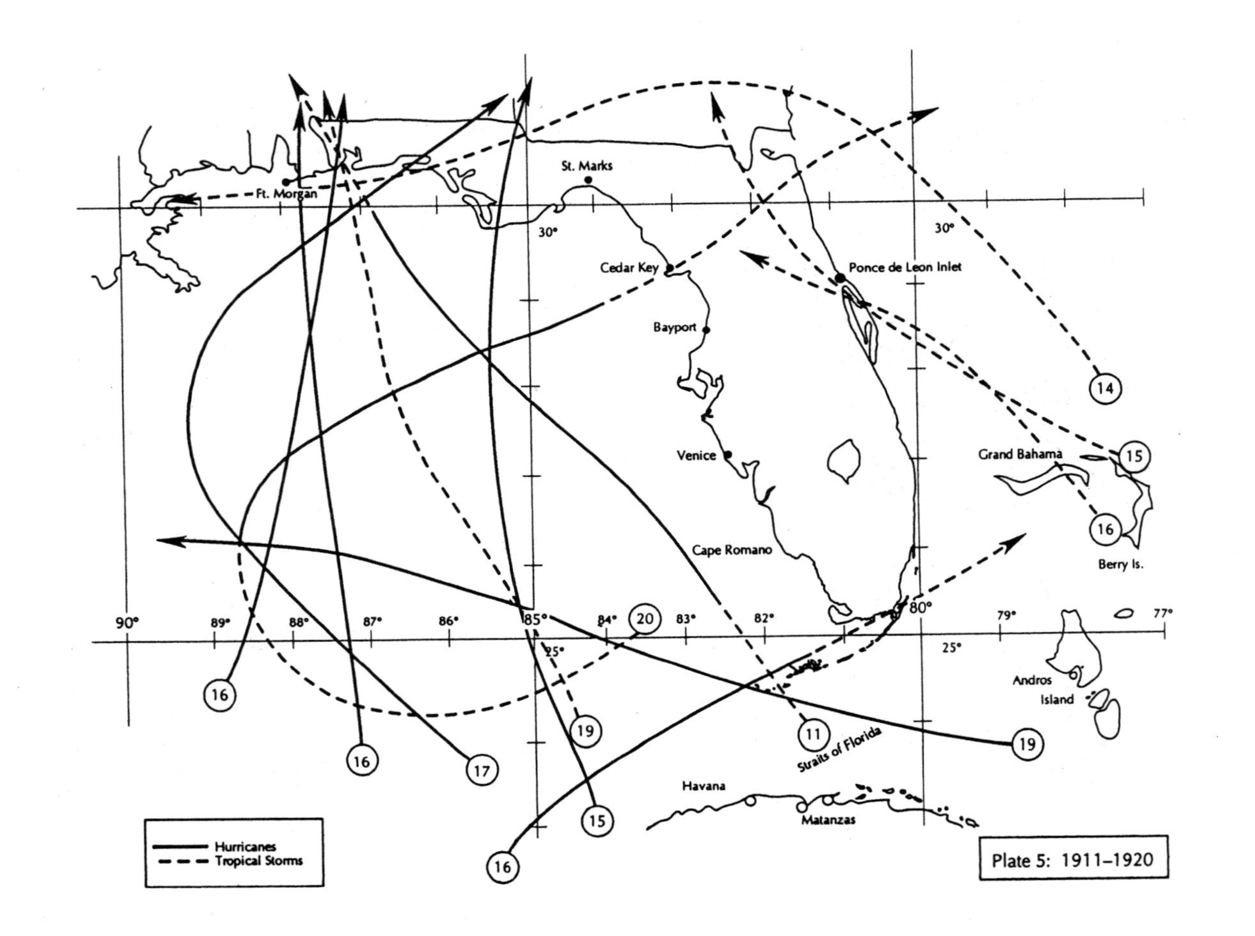
Ft. Morgan
St. Marks
30°
Cedar Key
Ponce de Leon Inlet
Bayport
Venice
Cape Romano
Grand Bahama
Berry Is.
Andros Island
Straits of Florida
Havana
Matanzas
90°
89°
88°
87°
86°
85°
84°
83°
82°
80°
79°
77°
25°
11
14
15
16
17
19
20
Hurricanes
Tropical Storms
Plate 5: 1911–1920

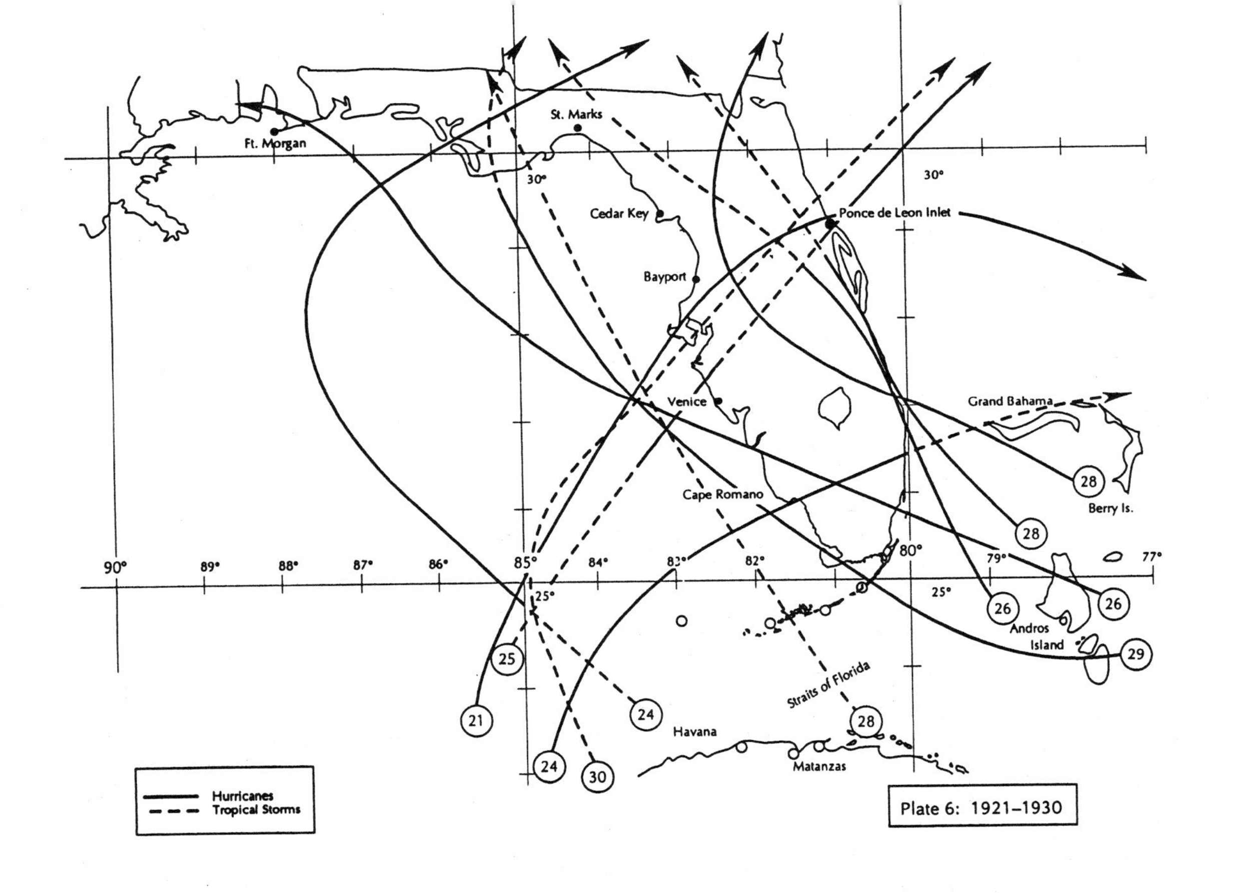

Plate 6: 1921–1930

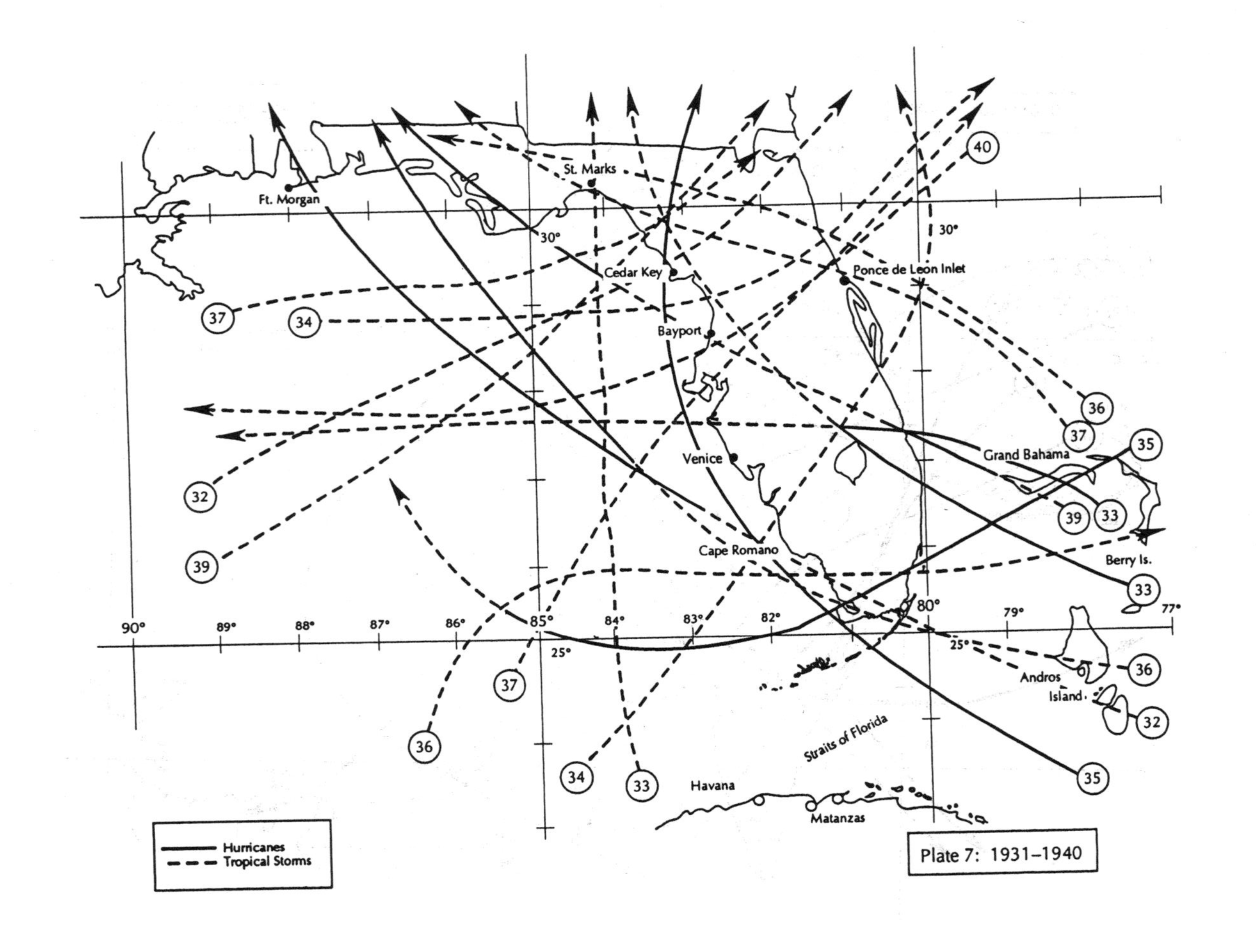

Plate 7: 1931–1940

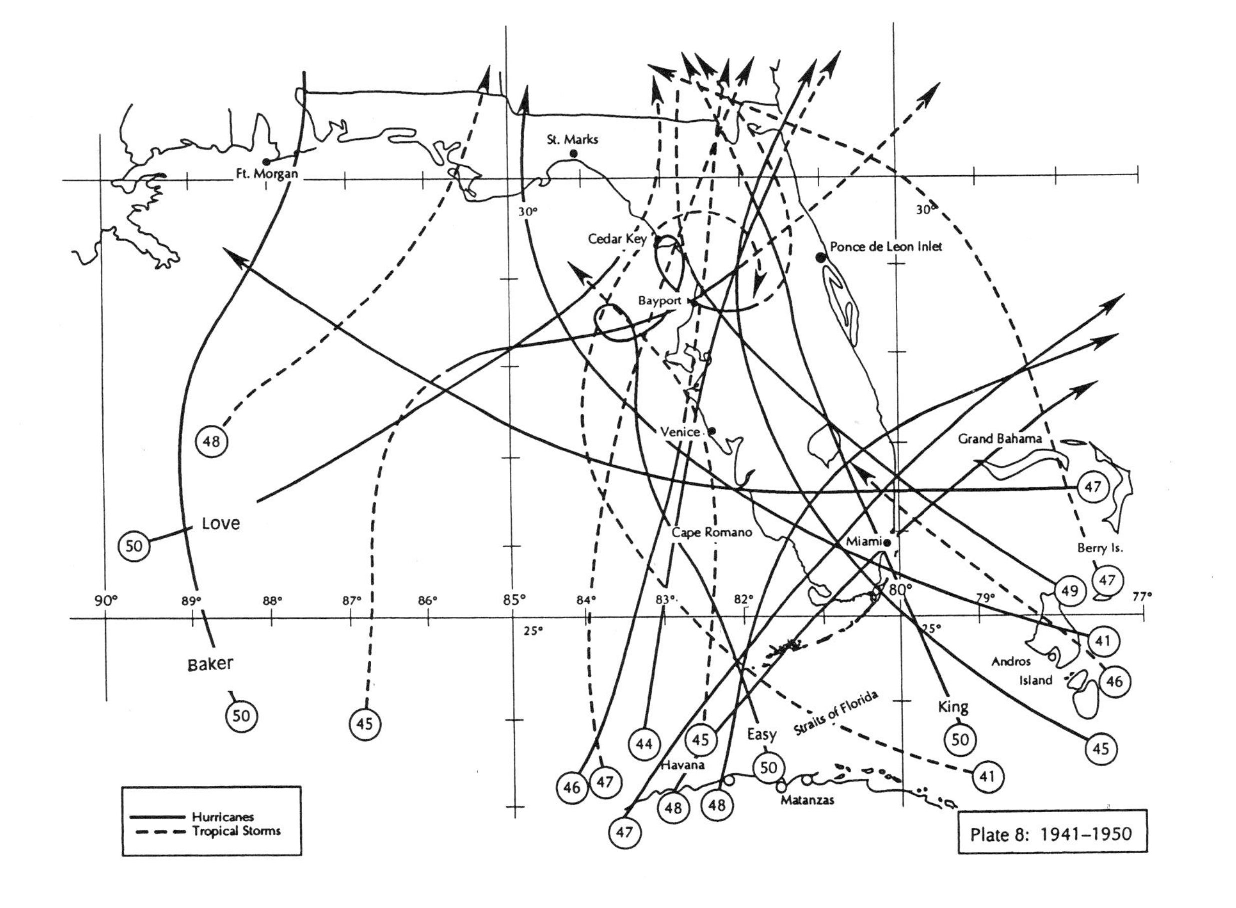

Plate 8: 1941–1950

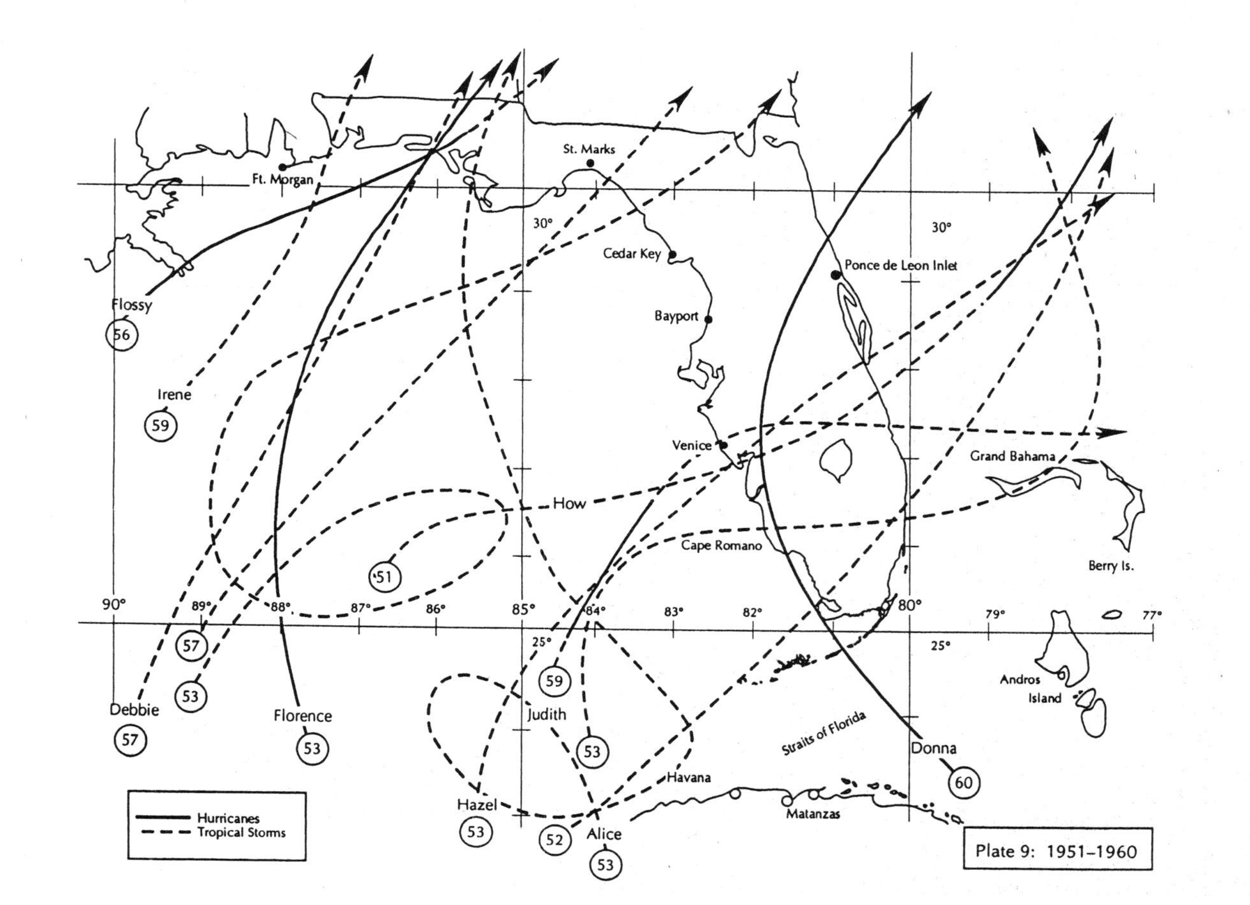

Plate 9: 1951–1960

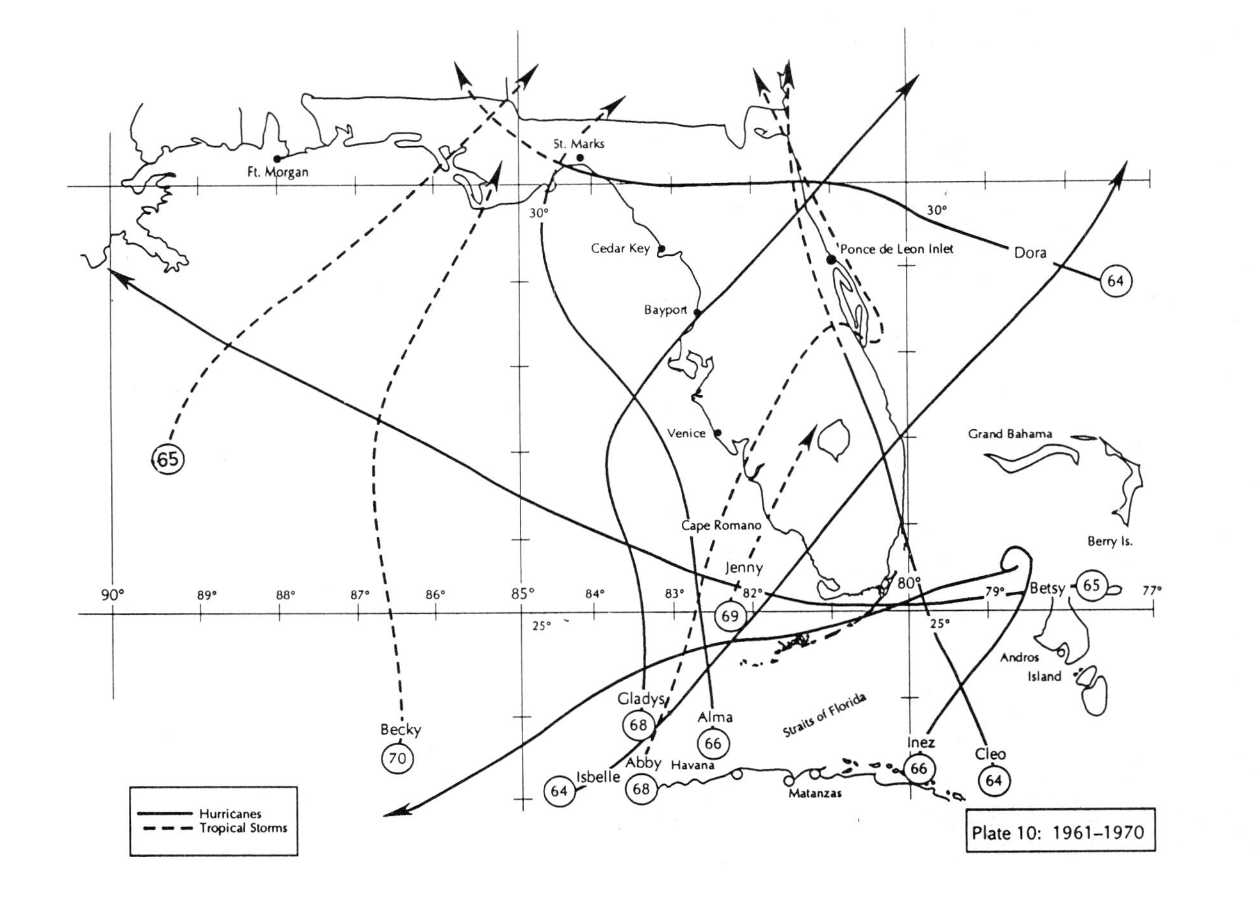
St. Marks
Ft. Morgan
30°
Cedar Key
Ponce de Leon Inlet
Dora
64
Bayport
Venice
Grand Bahama
65
Cape Romano
Berry Is.
Jenny
90°
89°
88°
87°
86°
85°
84°
83°
82°
80°
79°
Betsy
65
77°
69
25°
Andros
Island
Gladys
68
Alma
Straits of Florida
Becky
66
Inez
Cleo
70
Abby
Havana
66
Isbelle
64
64
68
Matanzas
Hurricanes
Tropical Storms
Plate 10: 1961–1970

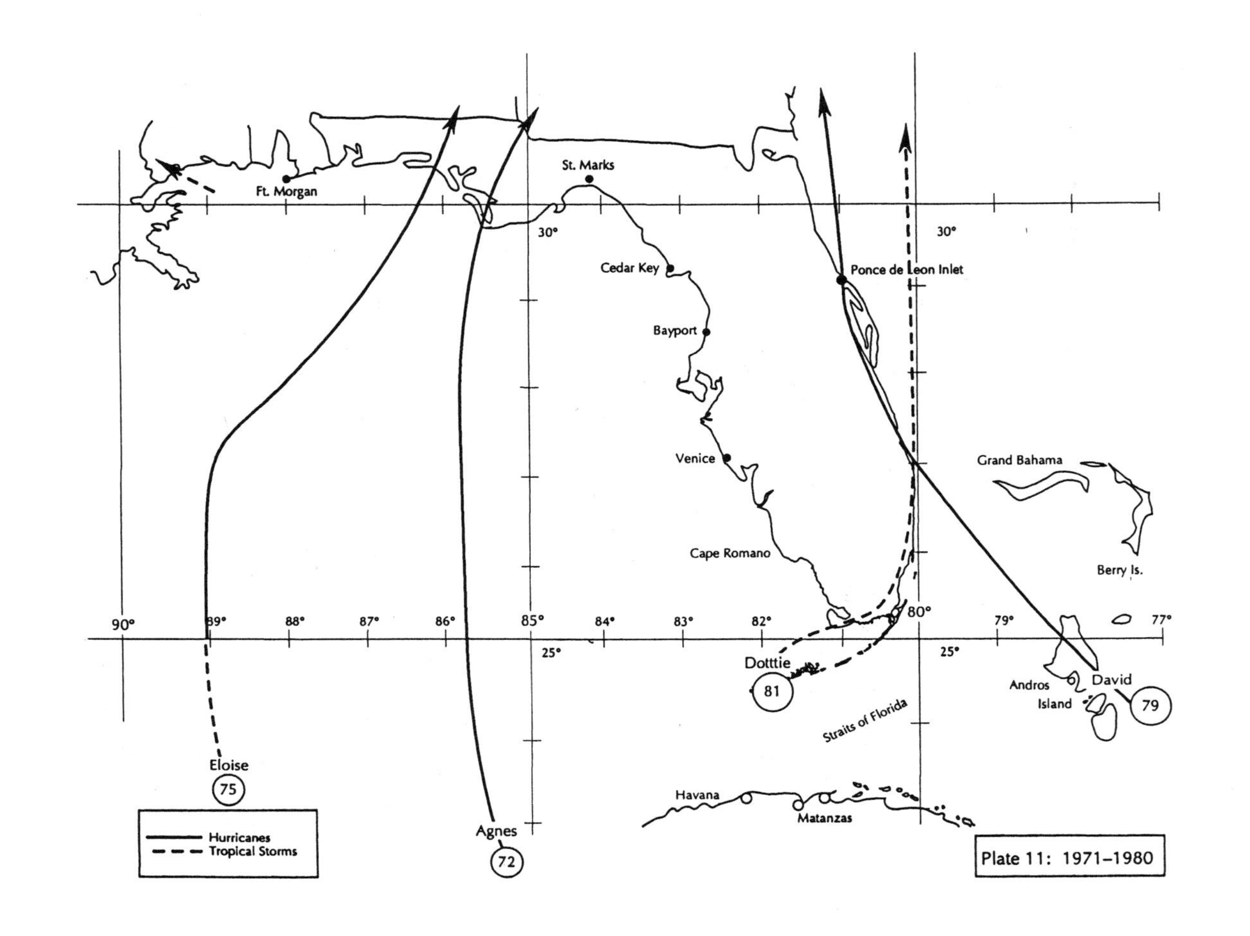
Ft. Morgan
St. Marks
30°
Cedar Key
Bayport
Venice
Cape Romano
Ponce de Leon Inlet
Grand Bahama
Berry Is.
90°
89°
88°
87°
86°
85°
84°
83°
82°
80°
79°
77°
25°
Dotttie
81
Andros
Island
David
79
Straits of Florida
Havana
Matanzas
Eloise
75
Agnes
72
Hurricanes
Tropical Storms
Plate 11: 1971–1980

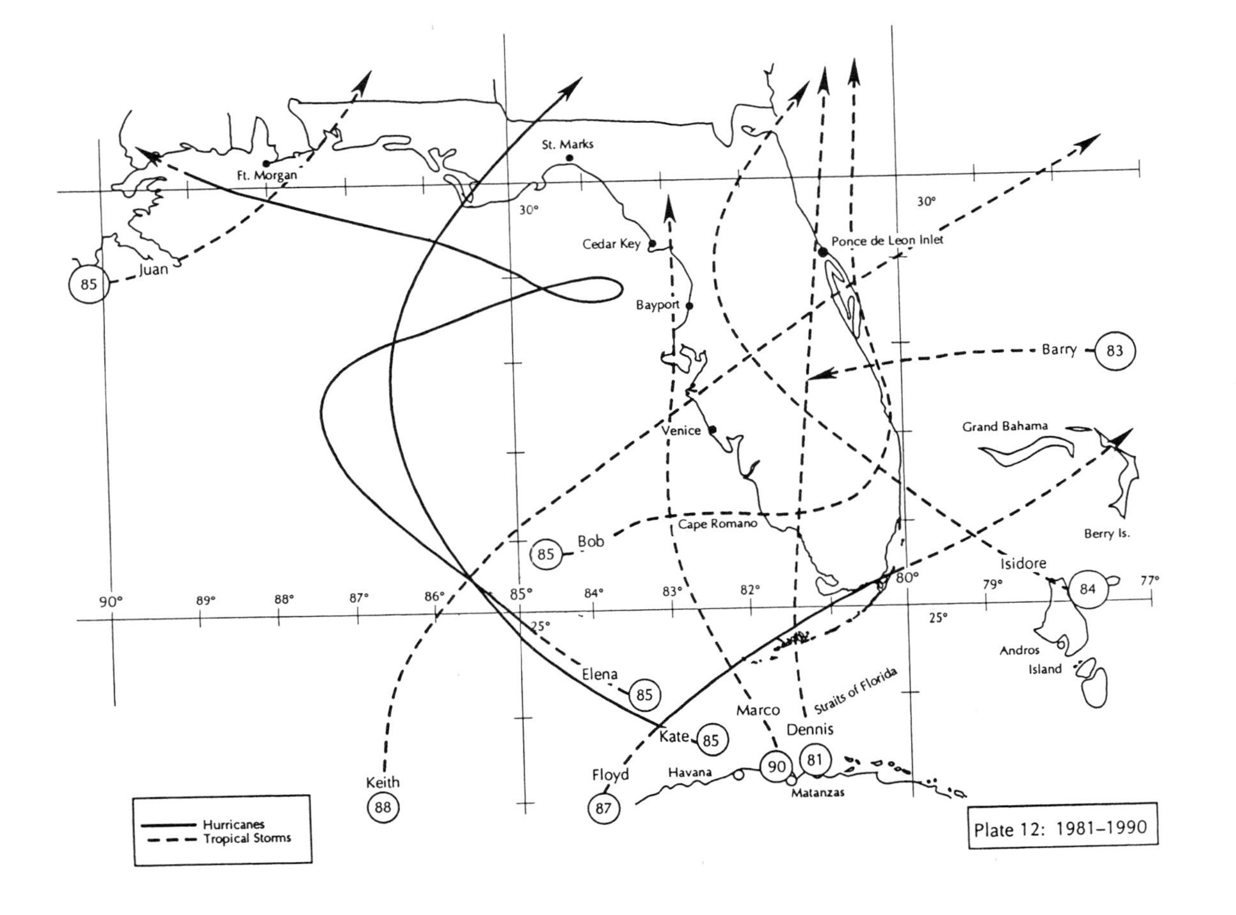
Ft. Morgan
St. Marks
30°
30°
Cedar Key
Ponce de Leon Inlet
Juan
85
Bayport
Barry
83
Venice
Grand Bahama
Cape Romano
Bob
85
Berry Is.
Isidore
84
90°
89°
88°
87°
86°
85°
84°
83°
82°
80°
79°
77°
25°
25°
Andros
Island
Elena
85
Marco
Straits of Florida
Dennis
Kate
85
90
81
Keith
88
Floyd
87
Havana
Matanzas
Hurricanes
Tropical Storms
Plate 12: 1981–1990

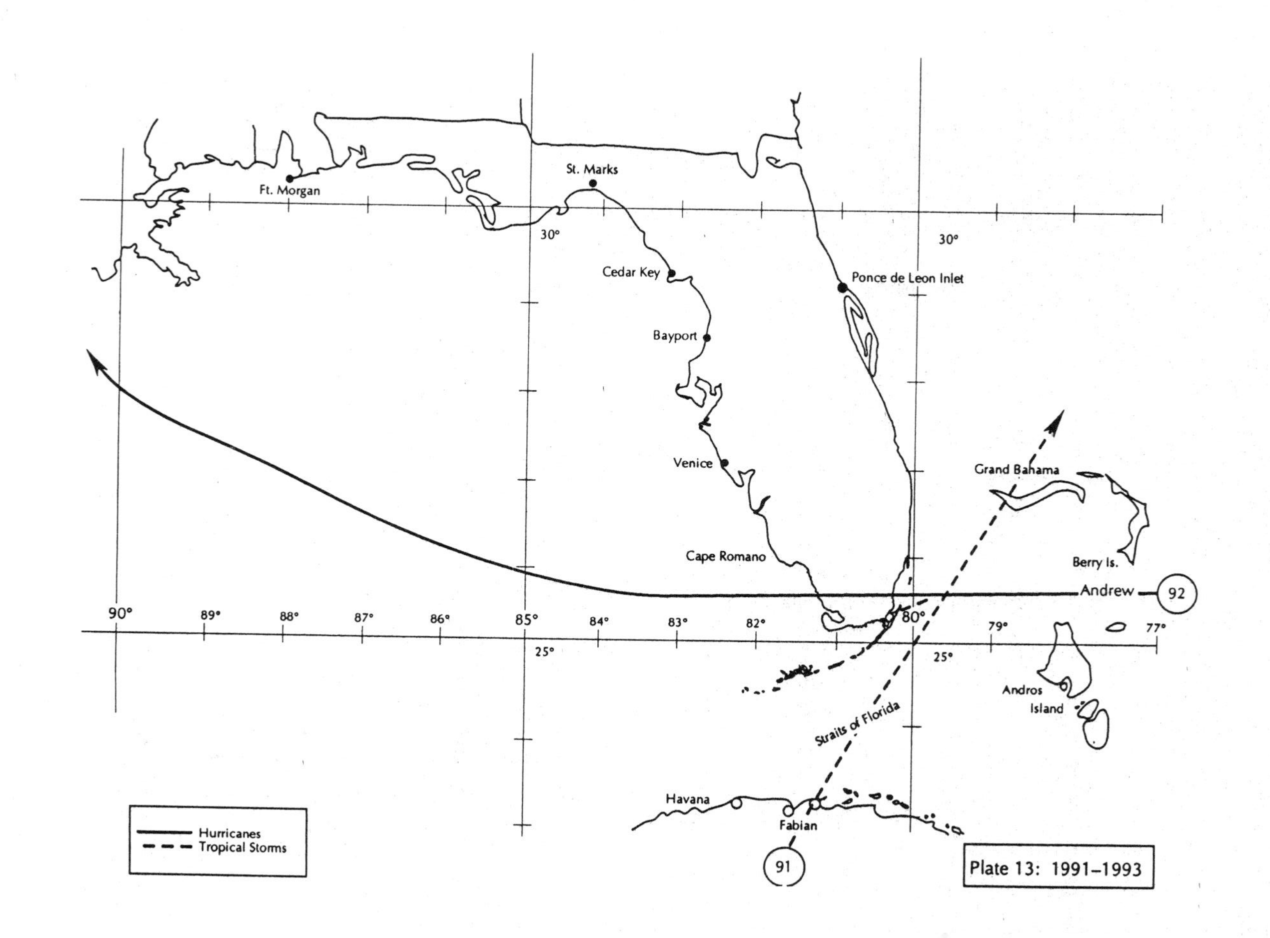

Plate 13: 1991–1993

Glossary[9]

1. **Tropical Cyclone**: By international agreement, Tropical Cyclone is the general term for all cyclone circulations originating over tropical waters, classified by form and intensity.

2. **Tropical Wave**: A trough of low pressure in the trade-wind easterlies.

3. **Tropical Distrubance**: A moving area of thrunderstorms in the Tropics that maintains its identity for 24 hours or more. A common phenomenon in the tropics.

4. **Tropical Depression**: A tropical cyclone in which the maximum sustained surface wind is 38 mph (33 knots) or less.

5. **Tropical Storm**: A tropical cyclone in which the maximum sustained surface wind ranges from 39-73 mph (34-63 knots) inclusive.

6. **Hurricane**: A tropical cyclone in which maximum sustained surface wind is 74 mph (64 knots) or greater.

7. **Tropical Storm Watch**: Is used for a coastal area when there is the threat of tropical storm conditions within 36 hours.

8. **Tropical Storm Warning**: A warning for tropical storm conditions, including sustained winds within the range of 39 to 73 mph (34 to 63 knots) which are expected in a specified coastal area within 24 hours or less.

9. **Hurricane Watch**: An announcement that hurricane conditions pose a possible threat to a specified coastal area within 36 hours.

10. **Hurricane Warning**: A warning that sustained winds of 74 mph (64 knots) or higher are expected in a specified coastal area within 24 hours or less.

---

[9] NOAA 1993.

11. **Storm Surge**: An abnormal rise of the sea along a shore as the result, primarily, of the winds of a storm.

12. **Flash Flood Watch**: Means that flash flood conditions are possible within the designated watch area - be alert.

13. **Flash Flood Warning**: Means a flash flood has been reported or is imminent - take immediate action.

14. **Small Craft Cautionary Statements**: When a tropical cyclone threatens a coastal area, small craft operators are advised to remain in port or not to venture into the open sea.

Hurricane Preparedness[10]

Be Prepared Before the Hurricane Season

- Know the storm surge history and elevation of your area.
- Learn safe routes inland.
- Learn location of official shelters.
- Review needs and working condition of emergency equipment, such as flashlights, battery-powered radios, etc.
- Ensure that enough non-perishable food and water supplies are on hand to last for *at least* 2 weeks.
- Obtain and store materials such as plywood and plastic, necessary to properly secure your home.
- Check home for loose and clogged rain gutters and downspouts.
- Keep trees and shrubbery trimmed. Cut weak branches and trees that could fall or bump against the house. When trimming, try to create a channel through the foliage to the center of the tree to allow for air flow.
- Determine where to move your boat in an emergency.
- Review your insurance policy to ensure it provides adequate coverage.
- Individuals with special needs should contact their local office of emergency management.
- For information and assistance with any of the above, contact your local National Weather Service office, emergency management office, or American Red Cross chapter.

When a Hurricane Watch is Issued

- Frequently monitor radio, TV, NOAA Weather Radio, or hurricane hotline telephone numbers for official bulletins of the storm's progress.
- Fuel and service family vehicles.
- Inspect and secure mobile home tie downs.
- Prepare to cover all window and door openings with shutters or other shielding materials.

---

[10] NOAA 1993.

Check Food and Water Supplies

- Have clean, air-tight containers on hand to store at least 2 weeks of drinking water (14 gallons per person).
- Stock up on canned provisions.
- Get a camping stove with fuel.
- Keep a small cooler with frozen gel packs handy for packing refrigerated items.
- Check prescription medicines - obtain at least 10 days to 2 weeks supply.
- Stock up on extra batteries for radios, flashlights and lanterns.
- Prepare to store and secure outdoor lawn furniture and other loose, lightweight objects, such as garbage cans, garden tools, potted plants, etc.
- Check and replenish first-aid supplies.
- Have an extra supply of cash on hand.

When a Hurricane Warning is issued

- Closely monitor radio, TV, NOAA Weather Radio, or hurricane hotline telephone numbers for official bulletins.
- Follow instructions issued by local officials. Leave immediately if ordered to do so.
- Complete preparation activities, such as putting up storm shutters, storing loose objects, etc.
- Evacuate areas that might be affected by storm surge flooding.
- If evacuating, leave early (if possible, in daylight).
- Leave mobile homes in any case.
- Notify neighbors and a family member outside of the warned area of your evacuation plans.

Evacuation

Plan to evacuate if you.....

- live in a mobile home. Do not stay in a mobile home under any circumstances. They are unsafe in high wind and/or hurricane conditions, no matter how well fastened to the ground.

- live on the coastline or on an offshore island, or live near a river or in a flood plain.
- live in a high rise. Hurricane winds are stronger at higher elevations. Glass doors and windows may be blown out of their casings and weaken the structure.
- stay with friends or relatives or at low-rise inland hotels or motels outside the flood zones. Leave early to avoid heavy traffic, roads blocked by early flood waters, and bridges impassible due to high winds.
- put food and water out for pets if you cannot take them with you. *Public shelters do not allow pets, nor do most motels/hotels.*
- Hurricane shelters will be available for people who have no other place to go. Shelters may be crowded and uncomfortable, with no privacy and no electricity. Do not leave your home for a shelter until government officials announce on radio and/or television that a particular shelter is open.

What to bring to a shelter:

- First aid kit; medicine; baby food and diapers; cards, games, books; toiletries; battery-powered radio; flashlight (per person); extra batteries; blankets or sleeping bags; identification, valuable papers (insurance), and cash.

If Staying in a Home:

- **Reminder**! Only stay in a home if you have not been ordered to leave. If you **ARE** told to leave, **do so immediately**.
- Store water - Fill sterilized jugs and bottles with water for a two week supply of drinking water. Fill bathtub and large containers with water for sanitary purposes.
- Turn refrigerator to maximum cold and open only when necessary.
- Turn off utilities if told to do so by authorities. Turn off propane tanks. Unplug small appliances.
- Stay inside a well constructed building. In structures, such as a home, examine the building and plan in advance what you will do if winds become strong. Strong winds can produce deadly missiles and structural failure. If winds become strong:
- Stay away from windows and doors, even if they are covered.

Take refuge in small interior room, closet, or hallway. Take a battery-powered radio, a NOAA Weather Radio, and flashlight with you to your place of refuge.

- Close all interior doors. Secure and brace external doors, particularly double inward opening doors and garage doors.
- If you are in a two-story house, go to an interior first-floor room or basement, such as a bathroom, closet or under the stairs.
- If you are in a multiple story building and away from the water, go to the first or seocnd floors and take refuge in the halls or other interior rooms, away from windows. Interior stairwells and the areas around elevator shafts are generally the strongest part of a building.
- Lie on the floor under tables or other sturdy objects.
- **Be alert for tornadoes which often are spawned by hurricanes.**

If the "EYE" of the hurricane should pass over your area, be aware that the improved weather conditions are temporary and that the storm conditions will return with winds coming from the opposite direction, sometimes in a period of just a few minutes.

AFTER the storm passes:

- Stay in your protected area until announcements are made on the radio or television that the dangerous winds have passed.
- If you have evacuated, do not return home until officials announce your area is ready. Remember, proof of residency may be required in order to re-enter evacuation areas.
- If your home or building has structural damage, do not enter until it is checked by officials.
- Avoid using candles and other open flames indoors.
- Beware of outdoor hazards - Avoid downed power lines and any water in which they may be lying. Be alert for poisonous snakes, often driven from their dens by high water. Beware of weakened bridges and washed out roads. Watch for weakened limbs on trees and/or damaged over-hanging structures.
- Do not use the telephone unless absolutely necessary. The system usually is jammed with calls during and after a hurricane.
- Guard against spoiled food. Use dry or canned food. Do not drink or prepare food with tap water until you are certain it is not contaminated.
- When cutting up fallen trees, use caution, especially if you use a chain saw. Serious injuries can occur when these powerful machines snap back or when the chain breaks.

Index of Named Hurricanes

*Not hurricane status in Florida.

Subject Index[11]

[11] name of hurricane or tropical storm underlined

Citation Index